Operations Management in Agriculture

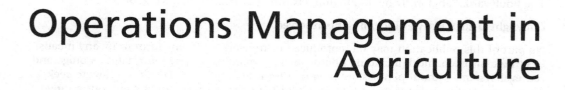

Operations Management in Agriculture

Dionysis Bochtis

Aarhus University, Department of Engineering
Aarhus, Denmark

Claus Aage Grøn Sørensen

Aarhus University, Department of Engineering
Aarhus, Denmark

Dimitrios Kateris

Centre for Research and Technology - Hellas
Institute for Bio-Economy and Agri-Technology
Volos, Greece

ACADEMIC PRESS

An imprint of Elsevier

Academic Press is an imprint of Elsevier
125 London Wall, London EC2Y 5AS, United Kingdom
525 B Street, Suite 1800, San Diego, CA 92101, United States
50 Hampshire Street, 5th Floor, Cambridge, MA 02139, United States
The Boulevard, Langford Lane, Kidlington, Oxford OX5 1GB, United Kingdom

Notices
Knowledge and best practice in this field are constantly changing. As new research and experience broaden our understanding, changes in research methods, professional practices, or medical treatment may become necessary.

Practitioners and researchers must always rely on their own experience and knowledge in evaluating and using any information, methods, compounds, or experiments described herein. In using such information or methods they should be mindful of their own safety and the safety of others, including parties for whom they have a professional responsibility.

To the fullest extent of the law, neither the Publisher nor the authors, contributors, or editors, assume any liability for any injury and/or damage to persons or property as a matter of products liability, negligence or otherwise, or from any use or operation of any methods, products, instructions, or ideas contained in the material herein.

Library of Congress Cataloging-in-Publication Data
A catalog record for this book is available from the Library of Congress

British Library Cataloguing-in-Publication Data
A catalogue record for this book is available from the British Library

ISBN: 978-0-12-809786-1

For information on all Academic Press publications visit our website at
https://www.elsevier.com/books-and-journals

Working together
to grow libraries in
developing countries

www.elsevier.com • www.bookaid.org

Publisher: Charlotte Cockle
Acquisition Editor: Nancy Maragioglio
Editorial Project Manager: Susan Ikeda
Production Project Manager: Denny Mansingh
Cover Designer: Matthew Limbert

Typeset by TNQ Technologies

Contents

Preface   ix

1. Agricultural Production Through Technological Evolution   1
   1.1 Key Phases in Agricultural Production Systems   1
   1.2 Production and Operations Management In Industry and Agriculture   14

2. Introduction to Engineering Management Basics   19
   2.1 Planning Level Definitions   21
   2.2 Project or Job Management Basics   30
   2.3 Designing and Organizing Production Systems   34
   2.4 Controlling Production Systems   39

3. Effectiveness and Efficiency of Agricultural Machinery   47
   3.1 Area Capacity   48
   3.2 Material Capacity   49
   3.3 Field Efficiency   49
   3.4 Machinery Systems' Productivity   73

4. Cost of Using Agricultural Machinery   79
   4.1 Direct Cost   79
   4.2 Indirect Cost   105

5. Choosing a Machinery System   117
   5.1 Tractor Selection   117
   5.2 Equipment Selection   128

5.3   Machinery Replacement   141

5.4   Machinery Management System Selection   148

6.   **Operations Management**   **159**

6.1   Optimization   160

6.2   Capacity Planning   161

6.3   Task Time Planning   162

6.4   Agricultural Vehicle Routing   169

6.5   Performance Evaluation   175

7.   **Agriproducts Supply Chain Operations**   **179**

7.1   Logistics Operations in Agricultural Production   179

7.2   Agrifood Supply Chain Management   181

7.3   Biomass Harvesting, Handling, and Transport Operations
      Management   182

8.   **Energy Inputs and Outputs in Agricultural Operations**   **187**

8.1   Energy Usage in Agriculture   187

8.2   Direct and Indirect Inputs   188

8.3   Assessment Tools   193

9.   **Advances and Future Trends in Agricultural Machinery and
     Management**   **197**

9.1   Robotics   197

9.2   Controlled-Traffic Farming   200

9.3   Precision Farming Management   200

9.4   Satellite Navigation   204

9.5   Farm Management Information Systems   205

10. Appendices   209
      Abbreviations   209

References   213
Index   223

Preface

The "green revolution" is a well-known term for the period between the 1930s and the 1960s and refers to the high number of technology transfers to agricultural production, including high-yielding varieties of cereals, artificial fertilizers, agrochemicals, and various new cultivation practices. The application of all these innovative technologies and practices was not possible without the mechanization of agricultural production. During the mechanization phase muscle-powered tasks were gradually taken over by machines, and agricultural production was transformed into an industrial-type system. Following that, and as a natural consequence, there was a period dedicated to rationalization of the work performed by workers and machines through effective management, in parallel with continuous increased replacement of manpower. During the most recent phase, as in all other application domains, the need for human sensory and mental input has been replaced by information and communication technologies (ICT) in association with automation systems, which overall has provided an increase in capacity, performance speed, and work repeatability.

During this evolution, the increased complexity of agricultural production systems caused by the increased complexity in the technical, economic, and social structures led to the adaption of methods and concepts of scientific management as measures to sustain effective and agile operations and production management. Agricultural operations management focuses on the design, planning, and operation of machine and human operations in agriculture. The objective is to ensure optimal planning and execution of operations in different agricultural production systems and supply chains. Targeted solutions are being implemented, using, among others, methods taken from other industries and branches and adapted to the unique agricultural working environment and domain.

Modern operations management in agriculture includes topics such as operations analysis, operations planning (e.g., mission planning, task time analysis, scheduling and allocation of resources), and operations optimization (e.g., capacity dimensioning, route optimization), among others. Furthermore, meeting the current trend of increased sustainability concerns in production systems, operations management tasks and processes must make the connection between decision-making and the corresponding environmental impacts.

This book aims at bridging the gap in the disseminated knowledge on recent advances in operations management in agriculture. It complements traditional aspects on a specific topic (e.g., cost of using and choosing machinery) with advanced engineering approaches that have been applied recently in agricultural machinery management (e.g., area coverage planning and sequential scheduling). In parallel, the book covers issues in the newly introduced technologies in bioproduction systems (e.g., robotics and ICT) and on the compliance of managerial tasks with environmental considerations.

Professor **Dionysis Bochtis** works in the area of systems engineering under enhanced ICT and automation technologies up to fully robotized systems. He has held positions as senior scientist in operations management at the Department of Engineering, Aarhus University, professor in agri-robotics at LIAT, University of Lincoln, and director of the Institute for Bio-Economy and Agri-Technology, CERTH.

Dr **Claus Aage Grøn Sørensen** is a Senior Scientist and the head of the operations management unit in the Department of Engineering at Aarhus University, Denmark. He holds a Ph.D. in production and operations management from DTU, Denmark. His research focuses on production and operations management, decision analysis, information modeling, system analysis, and simulation and modeling of technology application in agriculture.

Dr **Dimitrios Kateris** obtained a Ph.D. in agricultural engineering from Aristotle University of Thessaloniki. His research focuses on automation and new technologies in agricultural machinery, intelligent information systems in agriculture, and artificial intelligence. He is a senior researcher in the Institute for Bio-Economy and Agri-Technology, CERTH.

1

Agricultural Production Through Technological Evolution

1.1 Key Phases in Agricultural Production Systems

The "green revolution" is a well-known term for the period between the 1930s and the 1960s, referring to the number of technology transfers to agricultural production, including high-yielding varieties of cereals, artificial fertilizers, and agrochemicals and various new cultivation practices[1]. The application of all these initiatives was not possible without the mechanization of agricultural production. During the mechanization phase muscle-powered tasks were gradually taken over by machines, and agricultural production was transformed to an industrial-type system. Following that, and as a natural consequence, there was a period dedicated to rationalization of the management of the work performed by workers and machines, in parallel with continuous increased replacement of manpower. During the most recent phase, as in all other application domains, the need for human sensory and mental requirements has been replaced by information and communication technologies in association with automation systems, which provided an increase in capacity, performance speed, and work repeatability. Fig. 1 presents the time period of each of these phases. During this evolution a number of advantages emerged, including increased capacity (i.e., work performance), reduced labor cost and labor availability dependence, increased flexibility of the production system (easier adoption of new production practices), decreased material inputs (i.e., agrochemicals and fertilizers), and increased product quality (better control of processes). However, production systems became more complex, requiring higher investment and service costs (Fig. 2).

Operations Management in Agriculture. https://doi.org/10.1016/B978-0-12-809786-1.00001-1

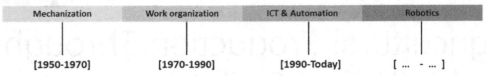

FIGURE 1 Agricultural production phases in terms of technology advances.

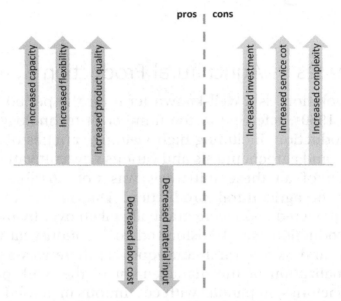

FIGURE 2 Advantages and disadvantages resulting from the technological evolution in agricultural production.

1.1.1 Mechanization Phase (1950–70)

In Europe the 1950s represent a very special period in the course of the modernization of agriculture, sometimes defined as a true agricultural revolution. In this period, agricultural machines were introduced to replace, for example, working horses and reduce by a large amount the use of labor. Mechanized production in comparison to muscle-powered methods has a number of benefits, such as:

- increased capacity, and thus increased cropped area
- reduced time for various operations
- independence of labor availability and seasonal labor shortages
- improved working environment and conditions for humans.

The introduction of mechanization in agriculture led to the implementation of industrial production methods, hence the organizational structures and management systems of industry also became an object of study in the agricultural domain. The model of economies of scale was adopted as a natural consequence. Economies of scale in agricultural production provide cost advantages due to the increased size of the operational environment (considering the field entity as the production "floor") and the processing units (machinery), and also the increased scale of operations. The reduction of the production unit cost derives from spreading out the fixed cost of the machinery (although increased in itself) to a higher number of units, and on the other hand on the reduction of the variable cost due to higher operational effectiveness as a consequence of the increased throughput capacity of the machinery and the reduction of various nonworking time elements (e.g., larger fields require less headland turnings).

However, there are physical limits in the economies of scales (Fig. 3). Some limits are generated by the desired flexibility of the system in terms of alternatives to either the production process (different farming system) or the final product itself (cultivation of another crop). Such changes require modifications to the machinery and consequently higher investment compared to a small production system. This is a critical point, especially for agricultural production systems where trends such as monoculture should be avoided. In agricultural machinery in particular, a physical limit is imposed by the increasing risk of soil compaction with increased machine size.

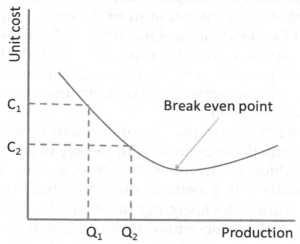

FIGURE 3 The general concept of economies of scales.

1.1.2 Work Organization Phase (1970—90)

In this period agricultural production more and more became an integral part of the general economic development of Western countries, and required new organizational structures to handle the increased complexity of the agricultural production system caused by complexity in its technical, economic, and social structures. As part of this adaption to new methods, the agricultural sector implemented the concept of scientific management within the production management domain, as initiated by Taylor and the Ford assembly line principles applied in industry.

Most ergonomic reviews and studies of the working environment included the analysis of accidents in agriculture as a consequence of using agricultural machines. These studies were initiated due to the increased levels of mechanization; key results indicated a significant correlation between an increase of the machine fleet and an increase in accidents. The analysis and evaluation of risk conditions in agricultural production systems provided valuable guidelines for machine manufacturers to improve their machines in terms of usage and safety.

The introduction of Taylorism in European agriculture had its beginnings in Germany, where studies were conducted in various companies and experience was acquired. This experience was diffused to other countries, such as the Netherlands, Switzerland, Poland, Belgium, and Finland. In France, Jean Piel-Desruisseaux was a key figure in the promotion of agricultural work organization based on Taylorism. In his famous book *Organisation scientifique du travail en agriculture*, written in 1948, he argues that every operation must undergo a foresight study to evaluate the need for personnel and machines. A foresight study predicts and obtains the optimal working conditions and disregards possible unfavorable working conditions. The first applications of operations research methods such as linear programming and simulation programs were also seen in this period.

During this time the concepts and application of work studies were accepted in an international context, with the key objective to assess and improve work productivity, decrease the physical stress of workers, and limit the risk of accident. In a methods study, the objective is to reduce the amount of work content by analyzing the whole work system and then identify possible options for removing inefficient operations. In work study measurements, the key objective is to identify, reduce, and finally remove nonproductive time.

Additionally, the application of work studies made it possible to derive standard task times for individual work elements within a whole work operation. Measuring times by using, for example, stopwatches is a key technique employed as part of a work study. Modern devices for task time measurements include personal computers (PCs) and IPads, and the latest technology developments make it possible to extract task times directly from the electronics on the applied machine. Normally, derived standard times for multiple work operations are stored in dedicated databases for further use.

The task time is the key variable in the work productivity estimation. The relationship between the measured task time and the standard time for a given task gives the work efficiency. The measurement and quantification of task times make it possible to estimate a number of operational parameters, like field capacity of an operating machine expressed in ha h^{-1} or acres h^{-1} (field capacity), or alternatively in t h^{-1} (material capacity for harvesters, etc.). The effective field capacity depends on factors like the working speed in km h^{-1} (or mph) and the working width of the machine in meters (or feet). The field efficiency is given by the relationship between the effective task time and the sum of the effective time and the nonproductive times which are considered. It may be expressed as:

$$FE = \frac{EFC}{TFC}$$

where FE is field efficiency, EFC is effective field capacity, and TFC is theoretical field capacity.

Field efficiency is influenced by a number of factors, including the size and the shape of fields, their width/length relationship, and the type of turning. Based on work studies from the late 1960s, manufacturers and advisors were able to derive practical support tools for estimating field capacity and other operational metrics without the use of, for example, PCs. These tools included nomographs capable of taking inputs like speed, width, and field efficiency to estimate effective field capacity (Fig. 4).

In the course of analyzing the factors affecting field efficiency, turning techniques and field size and shape were studied. Turning time is one of the nonproductive work elements which need to be minimized (Fig. 5).

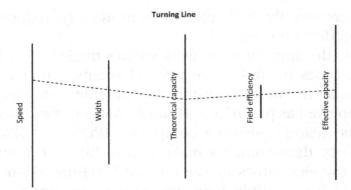

FIGURE 4 Topological nomograph for estimating field capacity.

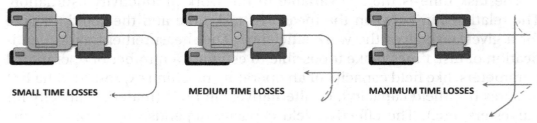

FIGURE 5 Effect of turns in operational time.

Turnings can be minimized but not eliminated completely. A number of different turning maneuvers were analyzed and evaluated: round, square, loop, and reverse corners, and loop and reverse turns. Next, the different types of turnings were connected to the various types of field shapes (Fig. 6).

Key results were also shown in terms of the relationship between field size and field shape as compared with the field efficiency, and as a function of working width and working speed (Fig. 7).

To analyze whole machine fleets, effective work capacities of machines were related to field areas that potentially could be worked by them. These studies included consideration of workability for specific operations (harvesting, seeding, etc.) depending on the weather and state of the soil. Based on these studies and estimations, it was possible to assess the performance of not only single machines but whole machine fleets. A general approach was developed in terms of analyzing and selecting machine fleets for whole farms. Key to this approach is the general work

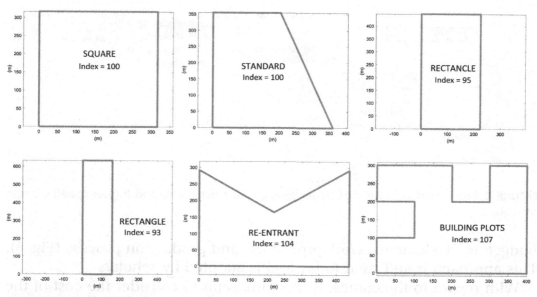

FIGURE 6 Effect of shape on an index of comparative time to complete a 10 ha field, square field index = 100^2.

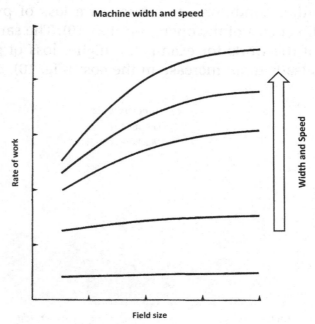

FIGURE 7 Quantitative representation of the combined effect of larger fields, wider machines, and higher speed on rate of work[2].

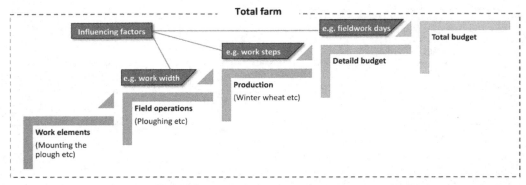

FIGURE 8 The combined effect of larger fields, wider machines, and higher speed on rate of work[2].

budget: work elements, work procedure, and production process (Fig. 8). This approach is still valid today and is adopted by scholars.

With regard to the choice, it was important to consider the cost of the work, including indirect costs related to punctuality and quality (Fig. 9).

The execution of an operation ahead of or behind schedule with respect to the time limitation, due to insufficient work capacity of the machine or unfavorable weather conditions, translates as a loss of production and therefore an indirect cost of the operation (Fig. 10). The same can be said for the quality of the work: for example, a higher loss of grain than the standard always leads to an increase in the cost (Fig. 10).

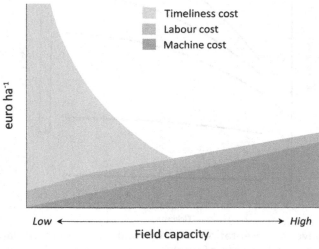

FIGURE 9 Total operational cost.

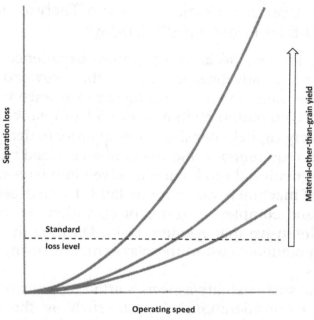

FIGURE 10 Quantitative representation of separation yield losses according to operating speed[3].

Advanced analytic tools were introduced at that period and computer-aided decision support systems. Also, a formalized mathematical modeling of the operations and related process (physical and technical) have been implemented. In this period, in parallel, the scientific area of energy in agriculture, in terms of energy consumption and the potential of producing renewable sources of energy originated from agriculture, was also related to the different levels of mechanization and the degree of the applied automation. On this basis, efforts were made to estimate the net energy equilibrium (the rate of output and energy flow in production) in various crops including crops for producing energy, food, and feed, taking into account direct (e.g. fuel consumption) and indirect (e.g. embodied energy of machines and materials such as chemicals and fertilizers) energy inputs.

In the 1980s the scientific area of computers and computational agricultural management, in terms of data analysis and decision making, became a new focus. That was the result of the increasing levels of mechanization and the appearance of automation-based applications as embedded systems of agricultural machines. That was the beginning of a new management system in agriculture, the well-known system of Precision Agriculture.

1.1.3 Information and Communication Technologies and Automation Phase (1990—Today)

From 1990 and for two decades agriculture experience and extensive transformation due to advances in genetics that provided new plant varieties with higher yield potential and higher resistance to diseases. The mechanical and automation technologies had to follow these advances and resulted in very capital intensive assets in order to deal with the large-scale production requirements and the new specialized types of farming. To this end, in agricultural production involved into large and specialized units, in terms of machinery and farming land. These expensive, in terms of investment, and complex, in terms of operations execution, systems required more intensive management tasks. On the other hand, the investment and operational cost of these systems was rising faster that the expected returns.

In parallel, the sustainability aspects started to be involved in agricultural production considerations. Aspects such as the environmental impact, food security, social impact, working conditions, and so on, have been considered and a number of regulations and standards have appeared for agricultural production to comply with. The planning focus was shifted to more completed tasks, e.g. scheduling, production strategies and organization of the farming system[4,5]. This new planning paradigm is focused on the ability of the system for automated decision making or decision support for the farmer (or the farm manager). The general notions of planning and scheduling took a central part on running farming. These tasks include the decomposition of any operation in a sequence of activities and, in a second step, the prioritizing of these activities in an optimal manner according to existing temporal constraints and defined optimal criteria. This change in the management paradigm generated the need for three key prerequisites. First, a systems engineering approach had to be implemented for fulfilling the analysis and integration of the various components. Second, advanced analytic management tools, such as tools for fleet management, logistics, scheduling, routing, and coordination, had to be introduced. Finally, in order to apply the above mentioned tools for data acquisition and data analysis also needed.

Table 1 Evolution of Operations and System Management in Agriculture

| Phases → | | | |
Entity ↓	**Manual Work Operation**	**Mechanization**	**Automation**
Process development	Experiences	Work studies	Models
	Method improvement	Method studies	System analysis
			System engineering
Planning	Routine	Work budget	Operations analysis
	Improvisation	Diagnostic models	Cost functions
		Time planning	Stochastic models
		Deterministic models	Prognostic models

As a consequence, the whole view on operations management in agriculture have been changed involving different entities in the different phases of agricultural development, namely the phases of manual work operations, mechanization, and automation. Table 1 lists the most important developments within agricultural operations management.

These developments of operations management through the various phases of agricultural production have been reflected in the scientific works of the last 20 years. There was a transformation from classical work science topics to engineering management topics, including themes such as operations research, data mining, data analysis, automation, and ICT applications. In parallel, in all the above topics sustainability and conservation aspects have been taken into consideration.

Finally, the introduction of robotics in agricultural production has also, once again, change the focus of the scientific approaches. New topics such as high-level control, distributed systems, machine learning, just to mention some, have received the interest of the relevant scientific community. All this research in agricultural robotics collaterally advances the research on conventional agricultural topics such as site specific management and precision agriculture processes.

It is worth noting that all these research applications had to deal with a wide range of operational environments, inducing:

- open environments (arable farming)
- semistructured environments (open-air horticulture), and

- controlled environments (greenhouses)., such as greenhouses, animal production, urban farming, processing plants for agrofood, wineries, etc.

As a natural consequence, a number of these advances had to be supported by web-based systems, either to provide support in the decision making based on the monitored data, or to automatically execute various activities. The adoption of the web-based applications provided a number of advantages[6]:

- standardization of methods and processes
- comparable results derived from different conditions
- available data for expert and non-expert users' considerations
- implementation of "cloud" services
- ability for easily evaluate different scenarios in terms of economic and sustainability impact
- availability of other scenarios generated by users elsewhere
- anonymity in provided data when this is imposed by personal or regulative requirements.

1.1.4   Agricultural holding definition

The term farm (or more officially "agricultural holding") refers to a single unit both technically and economically (implying a common use of labor and means of production, e.g., machinery, buildings, and land) which operates under a single management (independently if management activities are carried out by a single person, e.g., owner of the farm, or by a number of persons acting jointly) and undertakes agricultural primary or secondary activities, plus potentially providing supplementary nonagricultural products and services. An agricultural holding carries on at least one of the following activities for crop and animal production, hunting, and related services:

- growing nonperennial crops
- growing perennial crops
- plant propagation
- animal husbandry
- mixed farming

- holdings exclusively maintaining agricultural land in good agricultural and environmental condition.[1]

The number of agricultural holdings has been declining over recent decades. Based on EU statistics data, in the EU15[2] within a 12-year period (1995–2007) the number of holdings fell from 7.4 million to 5.7 million by (a decline of 23.2%), while in the EU27[3] within a 2-year period (2005–07) the number of holdings fell to 13.7 million (a decline of 8.8%). This trend is clearly depicted in Fig. 11. The reduction in agricultural holdings is related to two main factors, namely the land use change and the consolidation of small-sized holdings into larger-sized ones.

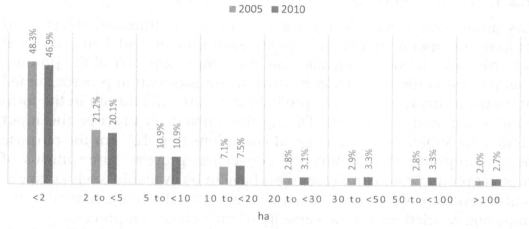

FIGURE 11 The distribution of agricultural holdings in the EU27 in size classes (ha) for 2005 (14,482 total number of holdings) and 2010 (12,015 total holdings). *Source: Eurostat.*

[1]Good agricultural and environmental conditions refer to a set of European Union standards and are directly related to issues such as maintaining the total area of permanent pasture, protection and management of water, soil erosion, soil organic matter, and soil structure.

[2]EU15 (1 January 1995–30 April 2004) comprised 15 countries: Austria, Belgium, Denmark, Finland, France, Germany, Greece, Ireland, Italy, Luxembourg, Netherlands, Portugal, Spain, Sweden, and United Kingdom.

[3]EU27 (1 January 2007–30 June 2013) comprised the EU15 countries and Cyprus, Czech Republic, Estonia, Hungary, Latvia, Lithuania, Malta, Poland, Slovakia, Slovenia, Bulgaria, and Romania.

1.2 Production and Operations Management In Industry and Agriculture

The generic task of the operations management function is the part of the total management task within the production system which has responsibility for "getting the job done". It is the objective of the operations function to provide the planning and control of the jobs and operations needed to produce the goods or services at hand. The importance of the industrial operations function is continuously increasing, as a challenge to improve quality, provide greater variety in the product or service, and meet customers' demands. The same increased focus on the operations function is being observed in farming.

1.2.1   Principles of the Production Process

Any production process involves a combination of humans, objects, and procedures operating within a given environment and being directed towards some predefined goals. The most important part of the production process is the conversion process, or transformation process, aimed at transforming resources and production factors into output in the form of products and/or services. During the conversion process, the input factors are subjected to a change of conditions (Fig. 12). For the purpose of initiating and maintaining the conversion process, some means of production (or production resources) have to be available, such as labour and machinery. Often these resources are available on a limited scale, imposing restrictions on the capacity of the conversion process.

Other important aspects involved in the production process are information and knowledge associated with the conversion process. Current

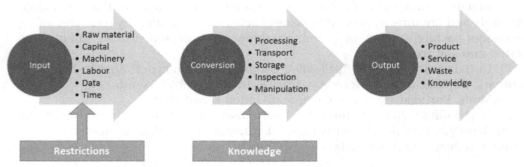

FIGURE 12 The general production process.

developments within all types of production processes require more and more knowledge-intensive activities to be undertaken and a comprehensive use of information technology. One of the most important objectives of modern operations management is knowledge acquisition, knowledge representation, reasoning, etc., as the basis for decision support. In general terms, the transformation process is the part of the production process, where the transformed material is given additional value related to potential consumers. This value can be added in the various elements of the conversion process, as depicted in Fig. 12.

- **Processing.** This normally indicates transformation of an entity in a physical sense. Examples include commodity-producing enterprises where raw materials are processed by means of casting, forming, mounting, etc. Within the agricultural system a material like mature grain in the field can be transformed by the work of harvesting into harvested grain in the grain tank of a combine and straw to be left on the field. Another example is the transformation of a stubble field into a plowed field through the work of a plowing operation.
- **Transport.** This indicates transportation of an entity from one location to another where its value is higher: some products are, for example, located at a retail sales site rather than in a factory. An example from agriculture is the transport of harvested grain from fields to storage facilities.
- **Storage.** This indicates the placement of an entity under certain conditions for a certain amount of time. Storage can either increase the value of an entity or uphold its value at a certain level. Within agriculture an example is the storage of harvested grain subjected to certain temperature limits.
- **Inspection.** Known also as registration, this may increase the value of an entity as its properties become more visible and recognizable. An example is quality control at a production line. Examination of a crop to check its health status or measure the moisture content of grain are examples from the agriculture domain.
- **Manipulation.** This indicates the combination or selection of raw data as comprehensible information to be used as the basis for planning and controlling operations.

The different standard types of transformation processes apply to both industrial and agricultural production. The precise version of the transformation process depends on the type of enterprise, but the generic characteristics remain identical. The next subsections outline in more detail the nature of the production process to reveal differences and similarities in industry and farming as related to the operations function.

1.2.2   Industrial Production Systems

Industrial production, in its normal sense, involves a production process from raw materials to finished products. However, the input to the transformation process does not need to be raw materials: it could be mere information or the customer himself/herself, as is the case for service production. Examples of processes where different kinds of input are used are listed below.

- Raw materials. A furniture manufacturer uses wood as an input, bringing it to a factory where it is cut and planed for polishing and assembling into a piece of furniture.
- Information. A tourist office collects and provides information to holidaymakers, assisting them in finding locations to stay and visit.
- Customers. In an airport, customers are the resources being processed from check-in to delivery at the awaiting plane.

One characteristic feature of these examples, and of all industrial production and service processes, is that the transformation is usually activated and controlled solely by human intervention. This is not to the same degree the case in farming.

1.2.3   Agricultural Production Systems

Agricultural production involves transformation that is realized by biological processes (e.g., crop biomass growth) taking place in the course of a growing season. The processes are regarded as an autonomous system which is basically independent of decisions made by the farmer. In contrast, interventions realized by labor and machinery during the plant-growing process are dependent on decisions made by the farmer and termed an *operation*. A formal definition of an operation was given by van Elderen[7]: "a technical coherent combination of treatments by which at a certain time a characteristic change of condition of an object (a field, a building, an equipment, a crop) is observed, realised or prevented". This definition extends operations beyond those for crop production to those supporting enterprise functions, like maintenance, repairs, etc. An operation is generally seen as the link between resources (e.g., labor and machinery), materials processed, and material produced (e.g., harvested crops, repaired machine, etc.).

Compared with the industrial production cycle, the difference is that most industrial processes are realized solely through human interventions initiated by decisions on how to transform raw material into produced goods, whereas the agricultural plant production cycle consists of an autonomous biological production process and a number of operations initiated by human interventions.

1.2.4   Agricultural Operations

Generally, agricultural machines comprise a number of components working together for a machine to perform the operation for which it is designed. For agricultural machines a number of basic processes exist, involving reversible, nonreversible, and unidirectional (neither reversible nor nonreversible) processes (Fig. 13).

When considering a specific agricultural plant production cycle, a number of cultural practices[8] are realized by executing a series of operations on soil or crops. These operations may for instance be soil treatment, seedbed preparation, seeding, fertilizing, plant care, irrigation, and harvesting. By executing *operations*, the objectives of the *cultural practices* are met. Table 2 gives a list of typical cultural practices present in crop production; for each process the nature of the operations and the associated type of transformation are outlined.

FIGURE 13 Basic work processes of agricultural machines.

Table 2 Typical Transformation Processes Within the Agricultural Plant Production System

Cultural Practice	Objective	Operation(s)	Transformation Processes	Types of Transformation[a]	Types of Treatments[b]
Primary tillage	Reduce soil strength, cover plant residues, rearrange aggregates	Observe soil status Plowing	Cut Pick-up Position Deposition	Inspection Processing	Observation Realization
Secondary tillage	Refine soil conditions prior to seeding	Observe soil status Seedbed harrowing	Dissociate Crush Grind Mix	Inspection Processing	Observation Realization
Seeding	Place seeds on or in the soil	Observe soil status Loading Transport Unloading Sowing	Pick-up Position Transport Meter Convey Position	Inspection Transport Processing	Observation Realization
Fertilizing	Provide nutrients for plant growth	Observe soil and crop status Loading Transport Unloading Spreading	Pick-up Position Transport Meter Convey Scatter	Inspection Transport Processing	Observation Realization
Plant care	Control weeds, insects, and plant diseases	Observe crop status Loading Transport Spreading	Pick-up Mix Transport Meter Convey Scatter	Inspection Transport Processing	Observation Realization
Grain harvesting	Recover grain from field, separate from rest of crop, move to storage	Observe crop status Combining Loading Transport Unloading	Cut Pick-up Separate Clean Convey Transport Store	Inspection Processing Transport Storage	Observation Realization

[a]Types of transformation as defined in an industrial context.
[b]Types of transformation (or treatment) as defined in an agricultural context.

2

Introduction to Engineering Management Basics

Any production process involves a combination of humans, objects, and procedures operating within a given environment and directed toward some predefined goals. All these are parts of the engineering management tasks within a production system (Fig. 14). The most important part of the production process is the transformation or conversion process, which transforms resources and production factors into output in the form of products and services. During this transformation, the input factors are subjected to a change of conditions. For the purpose of initiating and maintaining the transformation process, some means of production have to be available. These means (or production resources) include labor, machinery, etc. Often these resources will only be available on a limited scale, imposing restrictions on the capacity of the transformation process.

The operations system of an organization is defined as the set of all functions that produces the organization's products[9]; these can be either physical goods (tangible output) or services (intangible output). Operations systems consist of a series of elements that are common to all organization types. In the closed-loop operations system, these elements include the resources inputs, a conversion process (including a number of subprocesses of various types), the output of this process (a physical good or service), a feedback function that provides information on the function

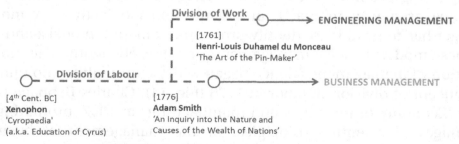

FIGURE 14 Evolution of management science.

Operations Management in Agriculture. https://doi.org/10.1016/B978-0-12-809786-1.00002-3

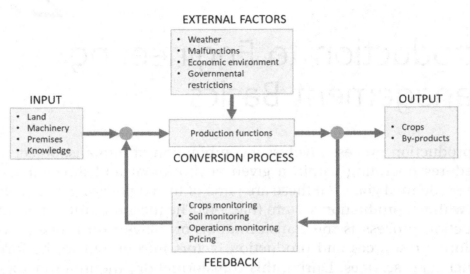

FIGURE 15 The operations system's elements in farming production.

execution, and a set of external factors (generally either unplanned or uncontrollable) that influence the actual output of the system and cause it to diverge from the planned one. Fig. 15 depicts the elements of the operations system in a farming organization in a closed-loop system notation.

To plan and control the conversion process, some kind of management is needed. Management may be defined as the art of "knowing what you want to do" and then seeing that it is done in the best and cheapest way. Management has always been a part of human evolution and undertakings, but with the work of Frederick W. Taylor it became a new scientific discipline for businesses and academia. Management science is a systematic and analytical approach to decision-making and problem solving using scientific methods of observation, experimentation, and inference as a way to reach an organization's objectives. A modern manager has to be able to identify problems, construct models, and then use these models to solve management problems efficiently. The modern evolution of management science began with Adam Smith, who outlined the concept of division of labor in 1776 (Fig. 14). Charles Babbage wrote about 'Economy of machines and manufacturer' in 1832, promoting the advantages of scientific principles for the management of business. Industrialization gave rise to serious problems in managing these new production systems, leading to, for example, F W Taylor advocating the development of a scientific management approach for planning and controlling work tasks.

2.1 Planning Level Definitions

The agricultural production processes within arable farming involve transformation that is realized by biological processes (e.g., crop biomass growth) taking place in the course of a growing season. The processes are regarded as an autonomous system, which is basically independent of decisions made by the farmer. In contrast, intervention realized by labor and machinery during the plant-growing process is dependent on decisions made by the farmer and termed an operation. A formal definition of an operation is given by van Elderen[7], who states that it is "a technical coherent combination of treatments by which at a certain time a characteristic change of condition of an object (a field, a building, an equipment, a crop) is observed, realised or prevented". This definition extends operations beyond those for crop production to supporting enterprise functions like maintenance, repairs, etc. An operation is generally seen as the link between resources (e.g., labor and machinery), materials processed, and material produced (e.g., harvested crops, repaired machines, etc.).

The complexity and importance of agricultural operations management have increased as agriculture has adopted capital-intensive production systems, thereby stimulating the development of more formal planning techniques. In addition, the trend toward sustainable farming practices has shifted the very nature of farm planning. From mainly dealing with the traditional way of planning what to do (which crops to grow and which machines to use), the focus has turned toward problems of how to schedule and carry out different work operations.

The generic term *planning* describes the process of reasoning about future events and the consequences of acting, choosing from among a set of possible courses of actions. Formally, a planning problem has as input a set of possible actions, a predictive model for the underlying dynamics, and a performance measure for evaluating the different actions. The solution to a planning problem is one or more actions that satisfy some specified performance requirements. Most planning problems are combinatorial, meaning that the number of possible actions or the time required to evaluate a given action is exponential in the description of the problem. Therefore, many planning problems can be shown to be NP-hard in terms of computational complexity. This complexity of planning problems often leads to the use of approximations and heuristic methods.

Another troublesome aspect is the inherent uncertainty and risk associated with agricultural operations. In a completely deterministic world it is possible to generate a sequence of actions knowing that if they are executed in the proper order the goal will be reached. However, in stochastic domains a planner must address the question of what to do when things do not go as planned.

The generation and execution of plans must be linked in a system monitoring effects of actions, unexpected events, and any new information that can lead to a validation, a refinement, or a reconsideration of the plan or goal. Plans must be presented in a conditional way, so that supplementary knowledge from observations, databases, sensors, and tests can be incorporated and integrated to revise the plan in the light of such new information.

Formally, the planning procedure makes use of different planning levels.

Strategic: Design of production system for a period of 1–5 years or two or more cropping cycles, and specifically the labor/machinery system in connection with the selected types of crops.

Tactical: Setting up a production plan for a period of 1–2 years or one to two cropping cycles narrowing down the resource usage, i.e., labor input and machinery input adjusted to the current crop plan.

Operational: Determining activities in the current cropping cycle, including short-term timing of activities and formulation of jobs and tasks.

Execution: Controlling the executed tasks and work-set performance.

Evaluation: Comparing planned and actual executed tasks.

The industrial production environment has comprehensively adopted the notion of formalized planning methods. It is the objective of planning to provide for the control of jobs and operations needed to produce the goods or services demanded by the customer. From being merely an internal way of controlling production efficiently (measured by levels of productivity), the planning and operations management functions are increasingly being recognized as a competitive weapon in the market[10]. This development clearly tends to increase the importance of operations management, as the challenge now is to use the function of operations management to improve quality, provide greater variety in the product or service, meet customers' demands as regards swift delivery, etc. Nowadays, a firm's competitiveness is to a large degree defined by its ability to deliver a certain product or service quickly and on time. The same increased focus on the operations function is being observed in arable farming.

In a generic sense, production and operations management are related to "activities, directed towards the preparation and execution of production processes in industrial enterprises and the development of management systems for planning and control of these activities"[11]. This overall definition without any modifications is applicable within an agricultural context. Differences between industrial and agricultural management are more likely to emerge when looking at the details for various planning levels with regard to the process of decision-making, planning environment, etc.

The basics of production management deal with what should be done, when should it be done, and by whom should it be done. The first and third questions determine which operations to execute and with what resources (process planning), while the second determines the times for executing the selected operations (scheduling). After the three steps have been considered, a reservation in the form <operation, time window, resources> is created. Decision-making regarding this reservation is at the centre of any scheduling problem.

The agricultural management task is different from the industrial one in a number of aspects, as for instance:

- the preponderant role of the environment, and the inherent uncertainty and risk (e.g., crop growth, weather conditions) characterizing any farm process
- the domain variables have relatively large variances
- the planning procedure has large time constants
- complexity in evaluating risky decisions

- agricultural operations can be halted by weather conditions, and consequently a job will be divided into part jobs.

To cope with this uncertainty and risk in terms of imperfection (imprecision, completeness) of available information and knowledge, any adopted arable farm management model has to fulfill the following demands:

- be flexible (i.e., tolerating that reality may be different from what has been anticipated)
- be adaptive (i.e., able to adjust to reality and the continuing flow of new information becoming available in the course of time)

Table 3 gives a description and structuring of information processing in the various planning stages in the two production types, industry and arable farming. Each information activity is described together with the information supplied and information required for this activity to be handled properly.

As indicated in the table, at the strategic planning level the focus is on the production dimensions and involves a static description or analysis of a given conversion process, where labor, equipment, and materials are maintained within given time and space specifications. Moving toward the tactical and operational planning level, the dynamic aspects of the conversion process come more and more into focus. Whether the dynamics influences the planning function depends on the type of production. In a traditional processing plant with a continuous flow of material the time aspect is not so important as when the output from the production process is controlled by varying customer demands. This is a so-called "job-shop"[12,13], which is defined as a type of production where the conversion process varies significantly as regards used materials, resources, process time, etc. In the case of continuous nonfluctuation process flow the time-dependent planning is to a large extent built into the conversion process at the design phase. The agricultural arable operation can be compared with the "job-shop" type of production process, where the material inputs, resources used, work time, etc. vary depending on the type of operation as well as being tailored to a fluctuating requirement for operations set by the weather and biological systems.

Table 3 Structure of Information Processing Within Strategic and Tactical Planning Functions

	Industry	Farming
Strategic planning or design of conversion system	Design of production system for a period of 1–5 years. Formulating targets and choosing organization structure. Information provided: • product specifications • process and method specifications • capacity specifications Information required: • aggregate product or services demand prognosis • possible equipment and facility layout • costs	Design of production system for a period of 1–5 years or two or more cropping cycles, specifically labor/machinery system + selection of types of crops. Information provided: • number and dimensions of machines • machine capacity • labor requirement • crops selected Information required: • possible production levels and price developments • operation demands • possible work methods • available machinery on the market • costs
Tactical planning	Planning of production size for a period of 3 months to 1 year. Information provided: • sales plan • production plan • inventory and job plan • material resource planning (MRP) • season leveling Information required: • strategic plan • sales prognosis • demand pattern (seasonal fluctuations)	Setting up a production plan for a period of 1 to 1–2 years or one or two cropping cycles. Information provided: • crop plan • machinery replacement • fertilizer/chemical application plans • maintenance plans • labor budget (peak loads) Information required: • strategic plan • availability of land, buildings, and equipment • external/internal standards

As in industry, strategic planning within arable farming concentrates on the development possibilities for the farm based on prices of production factors and sales of products. There can be assessments of possibilities for special or niche products, ecological production, transferring some parts of the production to other farms (equals "outsourcing" within industry), etc. Capacity planning concerns both qualitative and quantitative selection of production components (buildings, machinery, etc.) related to the demand. As in industry[12], the objective is generic optimization of the use of the components including their dimensions. This optimization is carried out by determining the following.

1. Demands of the operation to be performed.

2. Availability of equipment on the market.

3. Possible working methods.

4. Dimensions and capacity.

5. Costs.

6. Use of own machinery or contractors.

Many types of models supporting this optimization have been launched, from simple deterministic models[14] to more complex simulation and linear programming models. Characteristically the models work on the interaction between the labor and machinery system and the biological and meteorological system involving crops, soil, weather conditions, etc. The optimization considers constraints like available labor and machinery, timeliness, and workability. Compared to industry, capacity planning is performed within a domain characterized by a many uncertain factors relating to biological plant growth, weather, etc.

As is argued later, the strategic planning of machinery is interconnected with the operational planning. If this connection is not taken into consideration, a strategically chosen plan could turn up as not executable because it produces a nonworkable schedule.

Tactical planning within industry concerns the determination of the quantity and timing of production in the medium planning range. Plans are worked out for the size of production and implemented by way of plans for sale, inventory, jobs etc. The main objective is, within the boundaries of the strategic planning, to adapt and control production as regards a fluctuating demand. The method can be purely intuitive, but decision support in the form of mathematical models (e.g., linear programming, decision rules, simulation) is available.

In arable farming tactical planning involves setting up crop plans, fertilizer plans, feeding plans, and operations plans for the input of labor and machinery throughout the year. Normative estimation models for labor and machinery input given a specific crop plan and a specific machinery inventory are developed[15,16].

In industry, season leveling by use of inventory (for instance it is possible to produce inventory during idle periods) is a common practice. In arable farming the input of labor and machinery is completely constrained by crop development and weather, which is why peak load periods are present[17]. To reduce the organizational impact of these peak load periods, crops with different periods for intensive treatments have to be selected. Another thing is that generally the labor force in agriculture is flexible when it comes to undertaking the extra workload in these periods.

The tactical plans for production are further decomposed into short-term plans implementing and controlling the planned production activities (Table 4).

The distinction between operational planning and scheduling as part of establishing a sequence of actions complies with the artificial intelligence (AI) paradigm[18]. In the AI context planning concerns deciding whether to act and how (which acts to choose and in what sequence), while scheduling concentrates on timing of operations.

Operational planning is concerned with when a work operation can and may be executed to reach objectives like high delivery security, minimal process time, small inventories, and labor and machinery effectiveness. Short-term planning is especially tailored to order controlled production with varying demands.

With regard to the specific task of timing work operations, detailed job scheduling comes into play. In the agricultural context scheduling is defined as "determining the time when various operations are to be performed. Availability of time, labour and machinery supply, job priorities and crop requirements are some important factors"[19]. Work scheduling is the formulation of jobs based on required operations. Jobs can be scheduled when soil, crop, and weather conditions are within certain

Table 4   Structure of Information Processing and Flow Within the Operational Planning Function

	Industry	Farming
Operational planning	Master production scheduling and master capacity planning. Information provided: • output—capacity balancing • production schedules week by week Information required: • tactical production plan • customer orders • short-term demand forecasts	Determining activities in the coming cropping cycle, i.e., in the next few days to a year. The tactical plan is implemented by formulating, controlling, and adjusting an operational plan. Information provided: • required or optional operations • operations' urgency • operations' specifications Information required: • tactical production plan • internal/external standards • maintenance plan for land, buildings, and equipment

Scheduling	Detailed scheduling determining when to perform operations and for how long at each work center. Information provided:
	• start and finish times for operations
	• operation duration
	Information required:
	• operational plan
	• loading
	• sequencing
		Work scheduling setting up formulations of jobs. Planing implementation of work in the short term. Information provided:
		• work plan for planned operations indicating:
		  • start time
		  • duration
		  - work-sets required
		Information required:
		• required operations
		• urgency of operations
		• soil and crop status
		• weather forecast
		• workability criteria
		• availability of labor and equipment
		• operation specifications

limits. Planned jobs form the basis for task formulation (this could also be called implementation) (Table 5). Task formulation involves the actual specifications of work-sets performing the tasks, assigning instances to equipment items and worker categories making up work-sets.

Table 5 Information Processing Within Implementation, Execution and Evaluation

	Industry	Farming
Task formulation	Expediting in terms of tracking a job's progress. Information provided: • deviations from plans and schedules Information required: • equipment breakdowns • unavailable materials • priority changes	Handling tasks concerning inspection of formulated tasks. Information provided: • deviation from plans and schedules Information required: • equipment breakdowns • unavailable materials • change in soil, crop, or weather conditions • priority changes
Execution	Input–output control concerning work center performance. Information provided: • realized capacity • realized utilization Information required: • actual output • actual processing time	Controlling tasks and work-sets' performance. Task control Information provided: • realized work time • realized capacity Information required: • work time elements (effective time, ancillary time, preparation time, disturbance time, etc.) on work-sets Operation control Information provided: • set points for implement Information required: • operations' specifications Control device Information provided: • work quality • crop/soil conditions • realized yield • realized application • amount
Evaluation	Results of executed tasks and operations compared with internal/external standards.	Results of executed tasks and operations compared with internal/external standards

Execution involves different kinds of controlling the work being executed. Task control concerns the work efficiency of work-sets (measured by time elements and compared with standard data). Operation control concerns the transformation of operation specifications into set points for various parts of the equipment being used. Device control involves comparing machine operation (e.g., work quality) with planned specifications.

2.2 Project or Job Management Basics

In general, project management is the discipline of initiating, planning, executing, controlling, and finalizing specific activities to reach specific objectives and complying with specific constraints and criteria at a specified time. A project may be an undertaking to produce a specific product or service, or an integral part of an enterprise resource planning system. By this definition, the key element of project management becomes one of scheduling. Basically scheduling can be defined as "the allocation of limited resources to tasks over time"[20]. The solution to the scheduling problem involves assigning a starting time to every task. The feasibility of the schedule is measured by an objective like optimizing output from the process. Furthermore, several constraints must be met, such as the technologically fixed sequence of the tasks within a job, the capacity of a machine, etc. Operations are organized into jobs, and their order within a job is defined by technological requirements. The task of short-term scheduling is to select specific resources and exact times for all activities of the job.

Many scheduling techniques or models can be employed. The type chosen depends on the volume of work or activities to be scheduled, the nature of the activities, the overall complexity of the activities, and the extent of control the manager (project manager) want to exercise over the process. Basically, every scheduling problem consists of a number of elements and relationships, which any modeling effort must be able to capture.

Gantt introduced the well-known Gantt charts so the supervisor of a process is able to assess whether the process is on schedule, ahead of schedule, or behind schedule. Gantt charts are based on the simple principles that activities are measured by the amount of time required to complete them and the space on the chart (the bar length of a particular activity) represents the amount of the activity that should have been done in that time. Nowadays Gantt charts also show the precedence of relationships between the various activities.

In his 1916 book *Work, Wages, and Profits* Gantt[21] explicitly discusses scheduling, especially in the job-shop environment. He proposes giving to the foreman each day an "order of work" that is an ordered list of jobs to be done that day. Moreover, he discusses the need to coordinate activities to avoid "interferences". However, he warns that the most elegant schedules created by planning offices are useless if they are ignored, a situation that he observed.

2.2.1 The Modeling of Scheduling

The general scheduling problem identifies elements which are present in every scheduling problem. The different scopes of complexity in solving scheduling problems arise from differences in numbers of jobs, the way in which jobs arrive at the production unit, and the sequencing of jobs. Among others, Graves[22] has tried to classify the pool of scheduling problems according to certain criteria (Table 6). Production scheduling classification outlines a scheduling classification following the classification guidelines put forward by this author.

The general scheduling problem in a typical factory environment can be formulated as follows[23].

- Perform a number of jobs, each consisting of a given sequence of operations, by using a number of machines.
- To perform a job, each of its operations must be processed in the order given by the sequence.
- The processing of an operation requires the use of a particular machine for a given duration, the processing time of the operation.
- Each machine can process only one operation at a time.
- Given a cost function by which the cost of each possible solution can be measured, we want to find a processing order on each machine such that the corresponding cost and time requirements are minimized.

Different planning tools supporting the task of generating schedules have been developed, including Gantt charts, priority rules, critical ratio rules, etc[13]. Other scheduling methods are CPM (critical path method) and PERT (program evaluation and review technique)[24], which determine the most time-consuming path through a network of activities; the difference between the methods is that CPM is deterministic and PERT is stochastic

Table 6 Production Scheduling Classification

Classification Criteria	Classification Characteristics
Requirement generation or job generation	Scheduling requirement or jobs may be generated by either a customer demand process (open shop) or by way of inventory decisions (closed shop)
Requirement specifications	Requirement specifications may be termed deterministic or stochastic (e.g., processing time for each operation may be known or only specified by a probability distribution). The customer demand process may in a similar way be termed deterministic or stochastic.
Processing complexity or number of processing steps for each job	Different levels can be distinguished. • One stage, one processor, or the one-machine problem. All jobs require one processing step on one machine. • One stage, parallel processors. Each job requires one processing step performed on any of two machines. • Multistage, flow shop. Each job requires processing on a set of machines constrained within a given identical processing sequence. • Multistage, job-shop. All jobs require processing on a number of machines with no restrictions on the order of the processing steps.
Scheduling criteria or evaluation of schedules	Schedules may evaluated by schedule costs or schedule performance. Costs include production setup, variable production and operation costs, inventory costs, cost for not meeting deadlines, and costs for generation, implementation and monitoring of schedules. Performance may be measured by utilization levels for production resources, percentage of late jobs, average or maximum tardiness (difference between actual and desired completion time) for a set of jobs, average or maximum flow time (difference between completion and start time for a job).
Scheduling environment	Scheduling environments focus on assumptions about available information on future requirements. A static environment implies a finite set of fully specified requirements. A dynamic environment defines the scheduling problem in terms of known requirements as well as in respect of additional requirements and specifications generated over future periods.

with regards to time estimates. The methods provide a graphical display of project activities, an estimate of the project's time span, an indication of most time-critical activities, and an indication of how long an activity can be delayed without lengthening the project or activity.

Link and Bockup[25] stressed that the agricultural planning problem is unique, and well-known methods from industry, like CPM and PERT, do not directly apply in agriculture. Peart et al.[26] proposed the use of CPM/PERT models as activity networks for some applications of optimal machinery selection, but the models had difficulties in dealing with weather conditions, workability, etc.

Traditional operations research (OR) has tried to develop models (e.g., linear programming technique, dynamical programming technique) for different scheduling problems, with the intention of providing the planner with exact optimal solutions subject to a number of constraints[20,27]. Nonoptimal solution techniques like simulations have been developed in an attempt to provide a more flexible modeling approach.

According to the classifications in Table 6, the most complex scheduling problem are stochastic and dynamic, and that is exactly what most real-world situation are. Unfortunately, most scheduling models are deterministic and static. Consequently, only a few of these different scheduling models have been implemented as real-world applications. Pinedo[20] summarizes some reasons for this:

- often real-world scheduling problems are significantly different in scope compared to the mathematical models studied by researchers
- it is hard for practitioners to formulate their knowledge in the formal models
- the used operations research methods are complex and it is difficult to understand the results of such models.

The drawbacks of the OR methods stem from the fact that a rigid formalism is required, making it difficult to handle ill-structured knowledge. Overall, most scheduling models contain a wide variety of information that difficult to capture and represent, and dynamic changes in the scheduling environment require frequent rescheduling. Many authors state that the real problem is not scheduling but rather rescheduling[22,28]: it might be possible to determine a schedule, but the difficult part is the continuous schedule revision required by a stochastic and dynamic environment where information is uncertain or incomplete. In an industrial context, work on scheduling under uncertainty has been performed, but little has been implemented in practice.

As a consequence of these real-world scheduling characteristics, research within AI has tried to develop models overcoming the limitations of pure mathematical models. These efforts have led to the use of knowledge-based systems supporting but not replacing the human planner[29,30]. Among the best-known systems are ISIS[31], OPIS[32], and Micro-Boss[33]. The overall aim of these approaches is to develop more realistic models capturing and representing the uncertainties of the scheduling environment. Knowledge-based systems within AI differ from conventional computer programs in the way they solve problems. The central idea is to imitate the human reasoning process by relying on mixture of logic, belief, rule of thumb, opinion, and experience. In contrast, conventional models and programs solve problems by performing numerical calculations, solving equations, using optimization algorithms, etc.

2.3 Designing and Organizing Production Systems

This section describes the different aspects of the generic task of operations management (or production and operations management, as it is also often termed). Operations management is the part of the total management task within the production system which deals with the responsibility for "getting the job done". It is the objective of the operations management function to plan and control the jobs and operations needed to produce the goods or services demanded by the customer.

The basis for the following description is the operations management task found within either a commodity or a service-oriented firm. This description provides the foundation for a comparison to the principles of operations management in agriculture, specifically crop production. Differences and similarities are presented and used as guidelines when identifying the requirements for a model for the operational planning and control of field operations.

2.3.1   Principles of the Production Process

For the purpose of thoroughly understanding the operation management task, the basic characteristics of the production process and specifically the operations function are identified and outlined. Any production

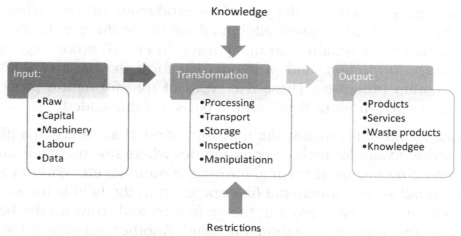

FIGURE 16 Production process[36].

process involves a collection of humans, objects, and procedures operating within a given environment and directed toward predefined goals. The production process transforms resources and production factors into output in the form of products and services. During this transformation, the input factors are subjected to a change of conditions (Fig. 16). The transformation process requires production resources (labor, machinery, etc.); often these resources are only available in limited amounts, restricting the capacity of the transformation process[34–36].

Another important aspect of production is the information and knowledge associated with a specific transformation process. The current trend in all types of production processes is that more and more knowledge-intensive work is being undertaken, requiring comprehensive use of information technology. Combined with human skills, this leads to the possibility of making more and more knowledge available to the decision-making process[37,38]. One of the most important objectives of modern operations management is knowledge acquisition, knowledge representation, reasoning, etc., as a basis for decision support.

It is important, in connection with the principle put forward here, to be aware that the transformation process is described in a general sense. The transformation in question can be a regular production process (the transformation of pure raw materials into final products), but often it does not imply a change in the physical composition of the material. In general terms the transformation is the part of the production process, where the transformed material is given additional value for the consumer. This added value can be obtained in many ways. In Fig. 16 *processing*, *transport*, *storage*, *inspection*, and *manipulation* form the basis of different transformation processes. These five types of transformation processes can be described in more detail on the basis of value added.

- *Processing* normally means the transformation of an entity in a physical sense. Examples include commodities where raw materials are processed by means of casting, forming, mounting, etc. Within the agricultural system a material like ripe grain in the field is transformed into harvested grain in the grain tank and straw on the field through the work of a combine harvester. Another example is the transformation of a stubble field into a plowed field by the plowing operation.
- *Transport* is the transportation of an entity from one location to another, assuming the value is higher when, for example, a product is located at a retail sales outlet rather than in the factory. An example from agriculture is the transport of harvested grain from the field to storage at the farm.
- *Storage* means the placement of an entity under certain conditions for a certain amount of time, e.g., storing foods in a cooler. Normally storage will increase the value of an entity, or at least uphold its value at a certain level. In agriculture, an example is the storage of harvested grain within certain temperature limits.
- *Inspection* or registration can increase the value of an entity as its properties become more visible and recognizable. An example is quality control on a production line. Examinations of a crop to check its health status or measure the moisture content of the grain are examples from agriculture.
- *Manipulation* means combining or selecting raw data to form meaningful information to use as the basis for planning and controlling operations.

As can be seen, the different types of transformation processes apply to both industrial and agricultural production. The precise transformation process depends on the type of enterprise, but the generic characteristics are universal. The following subsections outline in more detail the parts of industrial and agricultural production processes, with the intention of discovering differences and similarities in the operations function.

2.3.2 Industrial Production Systems

Industrial production its normal sense involves a production cycle from raw material to finished products. However, the inputs to the transformation process need not be raw materials, but could be information or the customer herself/himself in service industries.

- *Raw materials.* An example is a furniture manufacturer who uses wood as input; it goes into a factory, where it is cut, planed, polished, and assembled into a piece of furniture.
- *Information.* A tourist office gathers and provides information to holidaymakers, assisting them in finding places to stay and visit.
- *Customers.* In an airport, customers are among the resources being processed from check-in to delivery at the awaiting plane.

One characteristic feature of these examples, and of all industrial production or service processes, is that the operations function (or transformation process) is activated and controlled solely by human intervention. As will be seen. that is not the case in the agricultural production process.

2.3.3 Agricultural Production Systems

As discussed earlier, agricultural production processes in arable farming mainly involve transformation realized by biological processes (e.g., crop biomass, development stage, moisture content) in the course of a growing season. It is an autonomous system that does not depend on decisions made by the farmer.

Compared with the industrial production cycle, the difference is that the industrial process is realized solely by human interventions initiated by decisions, from raw material to produced goods, whereas the agricultural plant production cycle consists of an autonomous biological process and a number of operations initiated by human interventions.

2.3.4 Agricultural Operations

Generally speaking, agricultural machines comprise a number of components working together to perform the operation the machine is designed for. For agricultural machines, a number of basic processes exist[39], including eight reversible, four nonreversible, and three nondirectional processes. Table 7 lists these processes divided into the three categories.

Process diagrams containing parts formed by these different processes make it possible to describe any operation performed by agricultural machines.

Considering a specific agricultural plant production cycle, a number of *cultural practices* have to be realized[8] by executing a series of *operations* on a *soil* or *crop*. Examples of these operations are soil treatment, seedbed preparation, seeding, fertilizing, plant care, harvesting, and irrigation. By executing *operations*, the objectives of the *cultural practices* are met.

Identifying a crop management plan means setting up an ordered sequence of operations applied to a crop. For each crop grown there is a finite set of possible operations described by various attributes. Barley, for instance, involves seedbed preparation, sowing, fertilization, plant care if necessary, irrigation, and finally harvest. The task of designing a crop management plan involves choosing a subset of operations within the set of possible operations and defining how to execute each of these operations.

2.3.5 Describing and Defining Operations Management Activities on the Farm

In dealing with the operations function and different management models (like task time models, labor budgeting models, and planning

Table 7 Basic Work Processes of Agricultural Machines

Type of Process	Work Process
Reversible	Mix, fluff, pick-up, scatter, separate, pack, deposit, position
Nonreversible	Dissociate, cut, crush, grind
Nondirectional	Convey, meter, store

models) on the farm, there is a need for a structured formalization of management processes and information flows within the arable farm system. The information-handling activities, the required information for those activities, and the structure of the entities on which information is required can be summarized in a so-called *reference information model*[40]. Specific information models for arable farming and dairy farming have been created as guidelines when developing decision support systems for various parts of the farm system. The following subsections outline the most important data entities and information processes when analyzing and managing field operations. Firstly, in this regard the concept of "operations" is important. Table 8 defines the various terms and entities centered on the execution of farm operations.

The information associated with the operational activities of the farm is described in the attributes of the elements listed in Table 8. The attributes specify the kind of knowledge and data relevant to the decision-making processes connected to planning and implementing field operations on a farm.

2.4 Controlling Production Systems

In the previous sections, the transformation process and the scheduling problem were identified and described. However, the real challenge is to plan and control this process on the management level. A general model for the management of operations integrating different aspects is shown in Fig. 17. Centered on the conversion process, the subproblems of planning, organizing, and controlling are depicted. The framework outlined in Fig. 17 contains the elements worth considering when designing management systems by setting up models, analysis etc. The figure is generic in the sense that its overall structure can be applied to specific management processes.

In its most immediate form, control of a management process is based on the operations manager's intuition and experience, but often the manager will also use supplementary tools in the form of methodologies from management theories. These include classic, behavioral, and

Table 8 Structure of Arable Farm Operations

Cultural Practice	Cultural practices are seen as human intervention in the agricultural crop production system. To maintain production of crops grown under the influence and constraints of stochastic biological meteorological boundaries, a number of cultural practices have to be realized.[8] Examples include soil tillage, seeding, fertilizing, plant care, and harvesting.	**Attributes** • type
Operation	To realize the cultural practices the farm manager choose one or more operations, i.e., operations deal with *what* should be done to meet the objectives set by the cultural practices. The execution of a field operation transforms a material, like a crop or soil. Specifications like working speed, working depth, etc., must be given for each operation.	**Attributes** • operation type • work method type • job identification • specification for execution
Working method	The method by which the individual activities within an operation or chain of operations are carried out and coordinated with each other. A working method specifies *how* the operation is carried out in terms of type and number of operators and machinery, defining a gang[7] for that working method (e.g., operator1+combine harvester + operator2+trailer). For a gang to be formed, the availability of the labor and machinery items for that gang has to be considered.	**Attributes** • description of work method • sequence of activities • number and type of equipment • number of workers
Task	A task defines one or more operations carried out by a group of workers and machinery items, a work-set, working physically together (e.g., operator1 + combine harvester). Each work-set carries out one or more operations simultaneously or sequentially on a specified object following defined specifications and a certain working method.	**Attributes** • task type • operation type
Job	A job is one or more work-sets working organizationally together carrying out a combination of tasks. The work-sets work together for a certain period on some object following a certain working method. The combination of elements from the different work-sets forms a gang. In terms of work scheduling, the objective is to set up a schedule ensuring that all elements within the required gang are available during the planned period.	**Attributes** • job type • task type

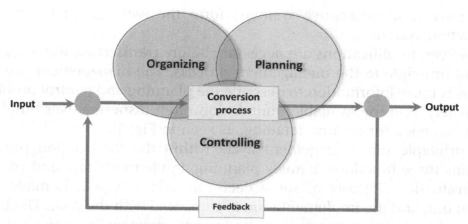

FIGURE 17 General model for managing operations[10].

modeling theories[10,41]. Each of these views emphasizes different aspects of the management task contained within operations management. The classic approach focuses on economical rationality, work efficiency, work division, scientific investigation of individual work operations, etc. So far, operations management in agriculture has for a great part been based on this classic approach. Behavioral management emerged later, and focuses on human relations and the environment in the production process. The modeling approach focuses on decision-making and systems theories, and how these can be modeled. The decision-making process is an integrated part of the management model, and as such it is vital to be able to model decision processes with an explicit representation of the relevant variables, so that a model can be built that is able to capture the possible consequences of different decision strategies.

A detailed management model can be built based on principles from the technical control process. The control of a technical process involves monitoring the output from the process: possible discrepancies from the planned output are reported and used as input in further controlling of the process. It has been shown that this control concept for a great part is comparable with planning and controlling processes at the higher management levels. In this context, the control concept is attached to the administrative units (controlling systems), which together with the productive units (executive systems) form the two main parts of any production system.

However, modifications are necessary before transferring the technical control principle to the management process. The management process demands more information to describe the planning and control problem completely, and the available information often disperses over time[13,42]. There is a need for greater detailing, as seen in Fig. 18.

In principle the management cycle within the information process contains three functions, namely planning, implementation, and control and evaluation. By way of some decision-making a plan is made and carried out, and the implementation is compared with the plan. The basis for decision-making processes is the overall objective for a given transformation process, types of decision rules, and registration data. In comparison with a simple technical control process, registration data has an extended meaning. Besides data concerning the actual processing, supplementary registration data from other systems or the environment are present; these indirect registration data are shown in Fig. 17. Inclusion of data from the environment helps in coping with the dispersion of information over time. Information from the environment will often take the form of a prognosis estimating state variables for future planning periods.

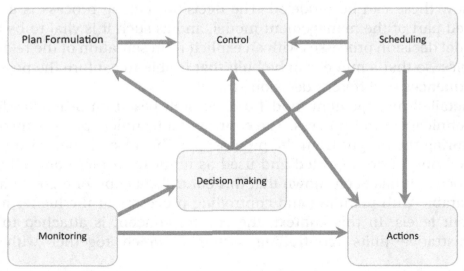

FIGURE 18 Planning and control of operational activities in the field.

The technical management task of any production system is to describe input, output, transformation, and production components for the production design and processing[43]. The production design concerns properties of the products (input, output), transformations, and components. Production processing concentrates on the actual execution of the process, determining the quantity and spatial and temporal setting for products, transformations, and components.

2.4.1 Managing Agricultural Operations

As with industrial management processes, the agricultural management process of planning, implementing, and controlling operations can be modeled, as shown in Fig. 18.

The decision-maker observes the status of the crop, field, weather, resources, etc., and given this information a decision is taken on whether to implement an operation or not. As time passes the planned operations are adjusted, because new information becomes available and operations get executed.

On the basis of a tactical plan, an operational plan is formulated for the following cropping period. Over time this plan is adjusted after observations of the crop and forecasts of, for instance, the weather, as well as the results of already executed operations. The implementation of the operational planning follows a set scheme.

- A planned operation is reported to the decision-making process. Based on its observed status, the decision-maker decides whether the operation is required or not.
- A required operation is reported to the scheduling process. This process coordinates the required operation with the other operational activities on the farm. On the basis of capacity, availability, and priorities, the labor and machinery process determines which operation to implement and at what time.
- The implemented operation is reported to the control/adjustment process for further planning and evaluation.

The operational planning process in agriculture is highly dynamic and interactive, putting a high demand on any proposed planning system. On top of that, there is only a sparse tradition of using formalized planning tools. That is in contrast to industry, where there is a long tradition of explicit planning schedules comprising formal documents passed down to the shop floor by the management for implementation. Farmers, on the other hand, generally both generate and execute any plan made, and their decision-making associated with planning remains very much implicit and external. That is similar to the managers of small businesses. The efforts to develop agricultural planning models must aim to externalize and formalize the planning effort, in this case the farmer's scheduling of field operations.

It should be noted that farmers' use of internal "mental" models also has advantages compared to computerized models: mental models are always at hand, have a certain element of adaptation to improve their performance over the course of time, and the inputs to the model are extensively based on the specific situation on the farm. The drawbacks are the lack of precision, especially in areas where the farmer has no experience. Using computer models to represent the domain and reasoning capabilities enables the farmer to base his/her decisions on a more comprehensive set of information of various types, gaining more precision in the decision-making process.

Specific differences between industrial and agricultural scheduling problems include the following.

- Agricultural scheduling requires the determination of a number of person/machinery complements, and not all technical equipment can be operated simultaneously.
- Agricultural operations can be halted by weather conditions, and consequently the job will be divided into part jobs.
- In agriculture some operations can be performed by the same human/machinery set, but the use of these sets is determined by a given sequence of operations.

To cope with this uncertainty and risk in terms of imperfection (imprecision, incompleteness) of available information and knowledge, any adopted agricultural management principle must be flexible and adaptive. Scheduling in arable farming has the greatest resemblance to management of the job-shop type of production in industry, and this is the scheduling method best suited to the characteristics of activity-based scheduling. The basic idea is that the activity most critical to the production is scheduled first, which is similar to the urgency concept in agriculture.

Planning in a dynamic and nonpredictive world like arable farming needs to be more robust than classic planning in more static and deterministic domains. Plans must be presented in a conditional way, such that supplementary knowledge from human or sensors observations and updated databases can be incorporated and integrated in order to provide a revised plan in the light of new information.

3

Effectiveness and Efficiency of Agricultural Machinery

Analysis of the efficiency of farm machinery involves a number of entities, such as job, operation, task, work element, etc., which have to be examined to assess the operational efficiency of labor and machinery on the farm. Table 10 explains the different elements as related to the example of the fertilizing-handling process. The "work element" functions as the basis for building models capable of estimating the labor requirement at different aggregated levels.

By adding the time content of the individual work elements outlined in Table 10, the result is the work requirement (or capacity) expressed as a function of a number of significant explanatory factors. The total labor requirement involved with work in fields may in principle be described by the following equation:

$$TRL = f(A) + f(B) + f(C) + \ldots f(X)$$

Table 10 Definition of Work Components

Designation	Work Operations: Field
Work element	Smallest activity unit identified, involving a sequence of working movements Example: "prepare for loading fertilizer tanker"
Operation	Sequence of work elements.[1] Example: "spreading organic fertilizer on the field"
Task	Sequence of operations. Example: "loading", "transport," and "spreading"
Job	A combination of tasks. Example: "fertilizer application"

[1]An operation is "a technical coherent combination of treatments by which at a certain time a characteristic change of condition of an object (e.g., a field, an equipment, a crop) is observed, realized or prevented."[7]

where *TLR* represent the total labor requirement (*TLR*) involved, and *A*, *B*, *C*, ...*X* each represent different part processes, e.g., plowing, harrowing, seeding, etc. The part processes may be described in a model as follows:

$$LP = f(x_1, \ldots, x_n)$$

where *LP* is the labor requirement involved with the part process and $x_1...x_n$ are the part operations comprising the part process. Some of these models are almost structurally identical, but have been adapted to individual part processes. Based on the task times, the area capacity, material capacity, and field efficiency may be estimated.

3.1 Area Capacity

The expected labor requirement and related machine capacity depend on many factors, such as traveling speed, working width, net capacity, yield, turning time, tending of machine, control, crop and soil stops, etc.

The labor requirement involved in field work can be calculated on the basis of various models describing the influencing factors determining the labor requirement and capacity. As an example, model *A* below can be used to describe all work processes involved with field work.

$$A = \frac{\left(\dfrac{h \cdot 600}{v \cdot e} + \dfrac{p \cdot b \cdot n}{e \cdot (1+a)} + k + s \cdot h\right)(I+q)}{h}$$

A = labor requirement, min ha^{-1}; h = size of field, ha; v = actual traveling speed, km h^{-1}; e = actual working width, m; p = minutes per turning; b = width of field, m; n = number of turnings per round; a = parameter, dependent on shape of field and traveling pattern; k = number of turnings on treatment of headland, min per field; s = crop and soil stops, adjustments, control, tending of machine, etc., min ha^{-1}; q = personal breaks, normally 5% additional time.

A is the labor requirement in min ha^{-1}, and the area capacity can be calculated as $A/60 \cdot h$ and expressed as ha h^{-1}. The traveling speed v is the mean actual traveling speed over the entire field; e represents the mean working width, that is the width of field or part of a field divided by an equivalent number of tracks with the implement in question; turnings p is the average time taken per turning; b is the overall width of field excluding lengthwise headland, if any, for instance in the case of plowing, sowing, mowing, etc.; the number of turnings n per round is specified for calculation of the total turning time; a is part of the calculation of number of turnings and equals 1 when traveling back and forth and traveling in circles on a rectangular field. In connection with tillage of the headland, the time of turning k is independent of the size of field, but dependent on shape of field, traveling pattern, and number of turnings; crop and soil stops and adjustments, control, and tending of machine (s), etc., are interruptions in the work and are assumed to depend on the area; personal breaks q are added as a percentage of the labor requirement.

The first phase in the equation is considered to be principal work, the next phase is turnings, and the remaining phases are described in the preceding paragraph. Only the principal work will increase the working results.

Table 11 gives examples of calculations of field work based on model A. The prerequisites used for the calculations are shown below.

Field size	4 ha
Field shape type	1:2
Transport speed on field	8–10 km h^{-1}

3.2 Material Capacity

The principal working time can also be calculated on the basis of the net capacity (t h^{-1}) of the machine and yield (t ha^{-1}), provided that the traveling speed and working width are not unreasonably large. In such cases the first phase in equation A can be replaced by the equation below:

$$y = \frac{h \cdot u \cdot 60}{d}$$

y = net labor requirement, min; h = field size, ha; u = yield, t ha^{-1}; d = machine net capacity, t h^{-1}.

3.3 Field Efficiency

The field efficiency factor depicts the overall rate of work to a theoretical spot rate of work. For example, typical values for the combine harvesting

Table 11 Examples From Calculation of Labor Requirement and Area Capacity for Some Implements Used in Soil Treatment

			Labor Requirement		
	Effective Working Width (m)	Traveling Speed (km h^{-1})	Field (min ha^{-1})	Total (h ha^{-1})	Area Capacity (ha h^{-1})
Four-furrow conventional plough	1.45	7.0	90	1.66	0.66
Three-furrow reversible plough	1.08	7.0	114	2.08	0.53
Six-furrow trailed plough	2.17	7.0	67	1.22	0.90
Five-furrow trailed/reversible plough	1.80	7.0	76	1.39	0.79
3.0 m heavy stubble cultivator, etc.	2.90	8.5	30	0.54	2.03
5.6 m seed bed harrow, etc.	5.40	8.5	15	0.28	3.94
6.0 m cambridge roller	5.80	7.5	17	0.32	3.49
3.1 m disc harrow	3.00	9.0	28	0.52	2.11
3.0 m rotary cultivator (5–7 cm)	2.90	6.5	41	0.75	1.46
2.5 m Finnish rotary harrow	2.40	11.0	27	0.50	2.19
7.6 m light harrow, 10-linked	7.30	8.0	13	0.24	4.65

process vary from 60 to 75 for a combine (Table 12). The magnitude of the nonproductive working elements of the harvesting operations is shown in Fig. 19 as the average relative percentage distribution of the part-time operations involved when harvesting with a combine harvester.

The variation in the value of the work elements depicted in Fig. 19 is caused by a number of factors:

- the theoretical capacity
- the size and shape of the field
- the operation patterns
- the maneuverability of the combine harvester
- the skill and experience of the combine operator
- crop conditions, like laid grain, weeds, etc.

The first part of model *A* describes the net work time of the machine and the other parts of the equation give the auxiliary time requirements shown, for example, in Fig. 19.

Table 12 Selected as Presented by ASABE standards[44]

Machine	Field Efficiency		Operating Speed	
	Range	Typical	Range	Typical
Tillage and Planting				
Moldboard plow	70–90	85	5–10	7
Field cultivator	70–90	85	8–13	11
Spring tooth harrow	70–90	85	8–13	11
Rotary hoe	70–85	80	13–22.5	19
Row crop cultivator	70–90	80	5–11	8
Row crop planter	50–75	65	6.5–11.0	9.0
Harvesting				
Combine	60–75	65	3.0–6.5	5.0
Mower	75–85	80	5.0–10.0	8.0
Mower (rotary)	75–90	80	8.0–19.0	11.0
Mower–conditioner	75–85	80	5.0–10.0	8.0
Mower–conditioner (rotary)	75–90	80	8.0–19.0	11.0
Windrower (Self-Propelled)	70–85	80	5.0–13.0	8.0
Side delivery rake	70–90	80	6.5–13.0	10.0
Rectangular baler	60–85	75	4.0–10.0	6.5
Large rectangular baler	70–90	80	6.5–13.0	8.0
Large round baler	55–75	65	5.0–13.0	8.0
Forage harvester	60–85	70	2.5–8.0	5.0
Sugar beet harvester	50–70	60	6.5–10.0	8.0
Potato harvester	55–70	60	2.5–6.5	4.0
Cotton picker (Self-Propelled)	60–75	70	3.0–6.0	4.5
Fertilizing–Spraying				
Fertilizer spreader	60–80	70	8.0–16.0	11.0
Boom-type sprayer	50–80	65	5.0–11.5	10.5
Air-carrier sprayer	55–70	60	3.0–8.0	5.0

3.3.1 Fieldwork Patterns

In vehicle motion studies, the (instant) turning radius is defined the distance between the instantaneous center of curvature (ICC) and a reference point on the vehicle. For example, in the case of a vehicle with a kinematic described by the bicycle model, the reference point is the center of the nonsteering

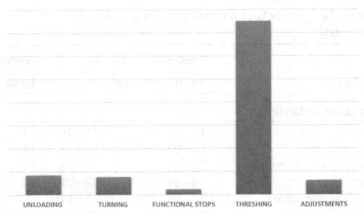

UNLOADING   TURNING   FUNCTIONAL STOPS   THRESHING   ADJUSTMENTS

FIGURE 19 Relative distribution of part-time work elements in the harvesting process.

wheel, while for a vehicle with a kinematic described by the tricycle model (Fig. 20 (left)) or the Ackerman steering model (Fig. 20 (right)), the reference point is the center of the axis of the nonsteering wheels. For the kinematic configurations described in Fig. 20, the relationships that give the minimum turning radius are as follows.

Tricycle model:

$$r_{min} = (O, ICC)_{min} = d \cdot tan(\pi/2 - a_{max})$$

Ackermann model:

$$r_{min} = (O, ICC)_{min} = d \cdot tan(\pi/2 - a_{1max}) - l/2 = d \cdot tan(\pi/2 - a_{2max}) + l/2$$

There are two factors affecting the lower bound of the minimum turning radius value. The first one concerns the kinematic restriction

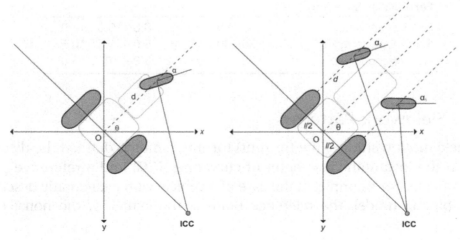

FIGURE 20 The tricycle and Ackermann steering models.

imposed by the maximum turn angle(s) of the steering wheel(s) that determines the minimum kinematic turning radius (r_{kmin}). The second factor concerns the restriction imposed by the maximum allowed value of the radial acceleration of the vehicle (a_{max}) for a given traveling speed (v), and determines the minimum dynamic turning radius[45] (r_{dmin}):

$$r_{dmin} = v/a_{max}^2$$

By taking these into account, the minimum turning radius is given by:

$$r_{min} = max\{r_{kmin}, r_{dmin}\}$$

The rough terrains where field operations take place and the soil conditions (in terms of soil texture and moisture content) affect the minimum dynamic turning radius of the vehicle. However, due to the low typical speeds during execution of headland turnings, this effect can be disregarded as a reasonable assumption, unless the moisture content is low and wheel slipping can occur.

3.3.1.1 Headland Turnings

To model geometrically the headland turning of an agricultural vehicle, the notion of Dubins curves[46] (or paths) must be implemented. "Dubins paths" are the shortest paths between two specified points with prescribed initial and terminal tangents, specific tangent directions, and bounded maximum curvature (or equivalent minimum turning radius). Dubins,[46] by means of geometric arguments, proved that shortest paths consist of a sequence of no more than three parts which are either a straight line (L) or a circle, denoted by C^+ when directed clockwise and C^- when directed counterclockwise. In total there are six types of Dubins paths:

$$C^-C^+C^-, C^+C^-C^+, C^-L\,C^-, C^+LC^+, C^+LC^-, C^-LC^+$$

The Dubins paths are extended by the Reeds–Shepp paths,[47] where traveling in the reverse direction is allowed. There are 48 different configurations in terms of "words" for the Reeds–Shepp paths. In a compact form that avoids the "+" and "−" notations, the different types can be summarized as:

$$C|C|C, CC|C, CSC, CC_u|C_uC, C|C_uC_u|C, C|C_{\pi/2}SC, C|C_{\pi/2}SC_{\pi/2}|C, C|CC, CSC_{\pi/2}|C$$

where symbol "$|$" denotes the reverse in the direction of motion, the subscript of "$\pi/2$" denotes that the specific curve must be followed for a length of $\pi/2$ radians, and u denotes that the parameters of another primitive must be matched.

Fig. 21 presents the results of the implementation of the Dubins and Reeds–Shepp algorithms for two-dimensional turnings. It can be seen that in the case of Dubins paths the traveled distance is considerably higher compared to the Reeds–Shepp paths for the same starting and ending points. However, the time requirements for the latter case are higher compared to the traveled distance due to change in the direction of motion.

Fig. 22 presents a number of headland turnings for an agricultural vehicle with a minimum turning radius of 5 m, corresponding to working widths of 4, 6, 8, and 10 m. It is clear that traveled distance during turning increases when the working width decreases. Fig. 23 presents the opposite case, where the working width is fixed and the minimum turning radius (actually the agricultural vehicles) is changed. It can be seen that for a fixed working width, the traveled distance during turning increases when

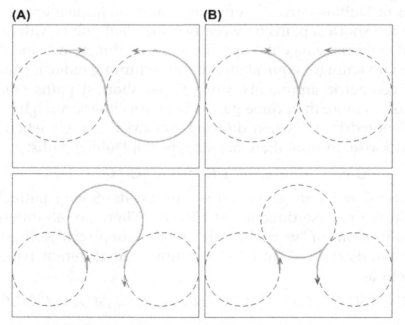

FIGURE 21 Two-dimensional vehicle turnings based on (A) Dubins paths and (B) Reeds–Shepp paths.

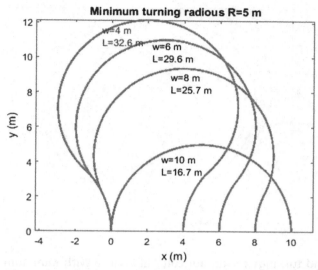

FIGURE 22 Headland turnings for an agricultural vehicle with minimum turning radius of 5 m and various operating widths.

the minimum turning radius increases. Figs. 24 and 25 present the same cases in terms of minimum turning radius and working with as Figs. 22 and 23, respectively, but for a field where the headland line has an inclination of 30 degrees to the perpendicular line of the field-work

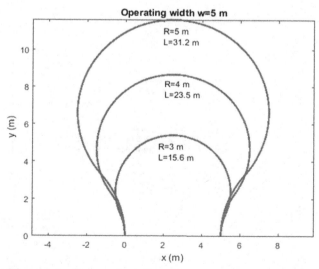

FIGURE 23 Headland turnings for agricultural vehicles of various minimum turning radius carrying equipment of 5 m operating width.

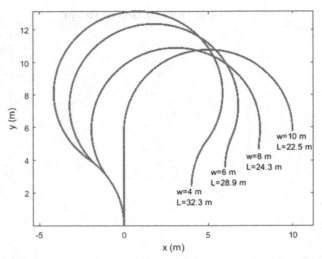

FIGURE 24 Headland turnings for an agricultural vehicle with minimum turning radius of 5 m, for various operating widths and a headland line inclination of 30 degrees.

track direction. Fig. 26 presents the effect of the minimum turning radius (as a feature of the agricultural vehicle) and the working width (as a feature of the carried equipment) to the traveled distance during a headland turning between two adjacent field-work tracks.

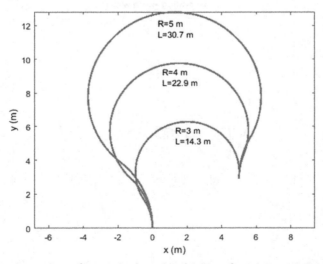

FIGURE 25 Headland turnings for agricultural vehicles of various minimum turning radius carrying equipment of 5 m operating width and a headland line inclination of 30 degrees.

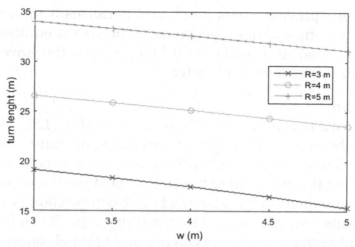

FIGURE 26 The effect of minimum turning radius and working width on the traveled distance during headland turnings.

3.3.2   Modeling Headland Turning

Fig. 27 presents the most common maneuvering types of headland turning executed by agricultural vehicles when the working direction has to be reversed. During an operation using the parallel field-work tracks system, the agricultural vehicle works by driving toward the upper maneuvering zone (upper headland); when reaching this zone it lifts the equipment section (e.g., disk harrow) or just interrupts the working

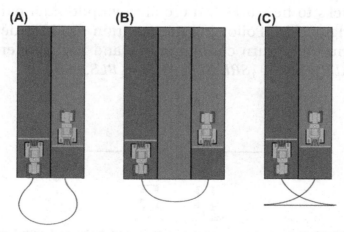

FIGURE 27 Headland turn types for agricultural vehicles: (A) Ω-turn, (B) Π-turn, and (C) T-turn.

process (e.g., in spraying operations), and performs a maneuver (headland turn); and at the end of maneuvering it lowers the equipment (or just restarts the process) and works by driving toward the lower headland, where the same procedure is repeated.

3.3.2.1   Π-Turn

The execution of a Π-turn requires the sequence $CSC \in \{LSL, RSR\}$ (Fig. 28). The Π-turn is the most comfort type of turn for the operator. However, to be feasible in execution it has to satisfy the restriction $x \geq r_{min}$, where x is the distance between the exit point of the current field-work track and the entry point in the subsequent field-work track. If this restriction is not satisfied, the operator must execute another maneuvering type. It can be considered that in terms of both operator convenience and optimal routing, the Π-turn maneuver is the "desirable" type of turning while the Ω-turn and T-turn are "constraint" types, in the sense that the latter two are performed only where a Π-turn cannot be performed.

3.3.2.2   Ω-Turn

A Ω-turn can be executed using any type of agricultural vehicles and carried equipment. Although the required time for the execution of a Ω-turn is less than that required for a T-turn, the Ω-turn has the disadvantage of higher requirements for available space in the headland area. As depicted in Fig. 29, there are various configurations of Ω-turn maneuvering and, depending on the relative positions of the current and subsequent field-work tracks to be worked, there are two potential part sequences for each configuration. Following the notation used for describing the parts of a turn, the Ω-turn configurations and part sequences include $CCC \in \{LRL, RLR\}$, $SCC \in \{SRL, SLR\}$, $CCS \in \{RLS, LRS\}$.

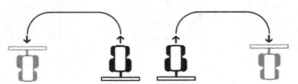

FIGURE 28 Π-turn in two configurations: *LSL* (left) and *RSR* (right).

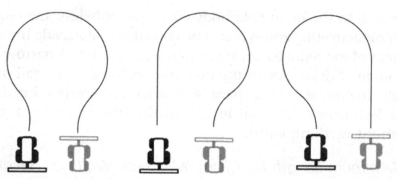

FIGURE 29 Ω-turn in three configurations: *CCC* configuration in *LRL* sequence (left); *SCC* configuration in *SRL* sequence (center); and *CCL* configuration in *LRS* sequence (right).

3.3.2.3   T-Turn

There are three different configuration of T-turn (Fig. 30): $C|S|C \in \{L|S|L, R|S|R\}, C|C|C \in \{L|R|L, R|L|R\}, S|C|C \in \{S|R|L, S|L|R\}$. The required time for a T-turn is considerably higher than that for a Ω-turn for the same pair of subsequent field-work tracks, due to the delay caused by reversing the vehicle direction of movement. For self-propelled machines the time requirement for execution of a T-turn is 30% higher than for case a Ω-turn, while in the case of trailed equipment the time requirements are doubled.[3] Nevertheless, the execution of a T-turn requires less free space and consequently a narrower headland area compared to the requirements for a Ω-turn.

From the agronomic point of view, the maneuvering zone is a "degraded" area in terms of its reduced productivity potential due to the increased soil compaction (increased machinery traffic during the headland turnings). Furthermore, there is an increased risk of the development of weeds and insect populations, as is generally the case in all field boundary

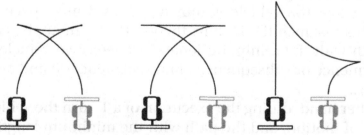

FIGURE 30 T-turn examples: configuration $C|C|C$, sequence $R|L|R$ (left); configuration $C|S|C$, sequence $R|C|R$ (center); configuration $S|C|C$, sequence $C|L|R$ (right).

areas compared to the main field body area.[3] Nevertheless, the requirement for reduced maneuvering zones has to be considered alongside the economic consequences of the reduced work rate of the machinery. A narrow headland area is a source of delay due to the execution of T-turns instead of Ω-turns. Traditionally, the maneuvering space is generated by considering the longest of either 1.5–2 times the total length of the tractor and equipment or the equipment working width.

3.3.2.4   Minimum Length of a Turn: Minimum Width of Headland Area

The length of a maneuver of a particular type and configuration depends on the kinematic characteristics of the machine and the distance between the two positions of the machine at the start and end of the maneuver. For maneuvers carried out when operating in a parallel field-work tracks system, this distance is always a multiple of the working width. Thus for agricultural work carried out by a specific vehicle (tractor or self-propelled machine) with a specific working width, the length of each maneuver type and configuration will be a function of this integral multiplier. To this end, we define as a maneuver "degree," denoted as d, the integer (positive) number that equals the quotient of the distance between the two subsequent field-work tracks and the working width of the machine. For example, if the machine after the end of one field-work track enters the immediate adjacent track, the maneuver degree equals 1, while if it leaves two field-work tracks unworked and enters the next one, the maneuver degree will be 3. The usefulness of this measure is shown in the next paragraph where its mathematical description is given.

In a case where the direction of motion of the agricultural vehicle remains the same during a complete turn (that is in a Π-turn and Ω-turn), it is considered that the vehicle is moving forward in a space free of obstacles. In this instance the Dubins theorem[46,48] can apply; according to this, the path with the minimum length between two vehicle configurations is a sequence or subsequence of line segments (S) and cycle arcs (C) of radius r_{min}.

On the other hand, during the execution of a T-turn the vehicle changes its direction of motion, and the path with the minimum length is derived by applying the Reeds–Shepp theorem.[47] According to this, the path with the minimum length can be described by a subset of the sequence $C|CSC|C$.

To simplify calculations, it is assumed that the configuration space (C-space) of the agricultural machine consists of 3° of freedom (x, y, θ), where x and y are the coordinates of the center of the rear wheel axle of the vehicle and θ represents the orientation of the vehicle. The actuation space is the radius of curvature of the course (r).

The integral and differential equations, respectively, which map the action space (A-space) of the machine in the C-space are:

$$\dot{x} = v \cdot \cos(\theta) \rightarrow x = x_0 + v \cdot \int_0^t \cos(\theta(t))dt$$

$$\dot{y} = v \cdot \sin(\theta) \rightarrow y = y_0 + v \cdot \int_0^t \sin(\theta(t))dt$$

$$\dot{\theta} = v \cdot r^{-1} \rightarrow \theta = \theta_0 + v \cdot t \cdot r^{-1}$$

where v denotes the linear velocity of the vehicle. According to the assumptions made, the action space (A-space) is a discrete space defined by the set $\{r_{min}, -r_{min}, \infty\}$. The set elements correspond to the right turn (r_{min}), the left turn $(-r_{min})$, and straight motion (∞).

The usual headland turns of an agricultural vehicle are completed in three steps. For example, during the execution of a Ω-turn to move the vehicle to a field-work track located at the right side of the vehicle as it drives upwards, the three steps are first, left turn $(L, r = -r_{min})$, second, a right turn $(L, r = r_{min})$, and third, a left turn again $(L, r = -r_{min})$. It is considered that during each step of the turn the velocity and turning angle are constant $(\dot{v} = \dot{\varphi} = 0)$. Now, let $\psi = [x, y, \theta]^T$ denote the state vector of the vehicle and Ψ the indefinite integral of the state vector. By integrating the differential equations stated above, for each one of the three steps $\psi_k = \psi_{k-1} + \Psi_k(t_k), k \in \{1, 2, 3\}$. By adding the three equations we then take $\psi_3 = \psi_0 + \sum_{k=1}^3 \Psi_k(t_k)$, where ψ_3 and ψ_0 are the initial and final state vectors, respectively. By solving the previous problem, the duration of each of the three steps is obtained. The total path length of the maneuver is computed by adding the corresponding line integrals, that is $\mu = \sum_k s(t_k)$, where:

$$s(t_k) = \int_c ds = \int_0^{t_k} \dot{r}(t)dr = \int_0^{t_k} \sqrt{[\dot{x}(t)]^2 + [\dot{y}(t)]^2}\, dt$$

For the maximum of the parametric function $x(t)$ one can obtain the width of the free maneuvering space required for the execution of the turn. To this number the quantity of $max\{w_v, w_e\}$ has to be added, where w_v is the width of the vehicle and w_e the width of the equipment.

The application of this method gives the functions to calculate the minimum length of each type (and configuration) of turn and the minimum width of the maneuvering zone to carry out each turn, as a function of the degree of maneuver for a particular agricultural machine (minimum turning radius) and a specific work (working width). These functions are listed in Tables 13 and 14, respectively.

Given that Ω and T types of maneuver are executed only when it is not possible to perform a Π-turn, a general function can be written that gives the distance that the agricultural vehicle has to cover during its move between two field-work tracks:

$$L_{min}(d) = \begin{cases} X(d), d > \dfrac{2r_{min}}{w}, X \in \{\Omega, T\} \\ \Pi(d), d \leq 2r_{min}/w \end{cases}$$

Table 13   Minimum Length Functions for Headland Turnings

Type	Configuration	Length Function
Ω	CCC	$\Omega_{min}^{CCC}(d) = r_{min}\left[3\pi - 4\sin^{-1}\left(\dfrac{2r_{min} + d \cdot w}{4r_{min}}\right)\right]$
	LCC, CCL	$\Omega_{min}^{LCC}(d) = \Omega_{min}^{CCL}(d) = r_{min}\left[\pi + 2\cos^{-1}\left(\dfrac{d \cdot w}{2r_{min}}\right) + 2\sin\left(\cos^{-1}\left(\dfrac{d \cdot w}{2r_{min}}\right)\right)\right]$
T	C|C|C	$T_{min}^{C\|C\|C}(d) = r_{min}\left[2\pi + \cos^{-1}\left(\dfrac{d \cdot w + 2}{4r_{min}}\right)\right]$
	C|S|C	$T_{min}^{C\|S\|C}(d) = r_{min}(2 + \pi) + d \cdot w$
Π	CSC	$\Pi_{min}^{CSC}(d) = r_{min}(\pi - 2) + d \cdot w$

Table 14   Minimum Required Headland Width

Type	Configuration	Length Function
Ω	CCC	$W_{min}^{CCC}(d) = r_{min}\left[1 + 2\cos\left(\sin^{-1}\left(\frac{2r_{min}+d\cdot w}{4r_{min}}\right)\right)\right] + \max\{w_v, w_e\}$
	LCC, CCL	$W_{min}^{LCC}(d) = W_{min}^{CCL}(d) = 2r_{min}\cos\left(\sin^{-1}\left(\frac{2r_{min}+d\cdot w}{4r_{min}}\right)\right) + \max\{w_v, w_e\}$
T	C|C|C	$W_{min}^{C\|C\|C}(d) = (r_{min} + \max\{w_v, w_e\})\sin\left[\cos^{-1}\left(\frac{d\cdot w + 2r_{min}}{4r_{min}}\right)\right]$
	C|S|C	$W_{min}^{C\|S\|C}(d) = r_{min} + \max\{w_v, w_e\}$
Π	CSC	$W_{min}^{CSC}(d) = r_{min} + \max\{w_v, w_e\}$

It should be noted that the length provided from the function above is theoretical, and its calculation is based on the assumptions mentioned earlier. This corresponds to the minimum traveled distance during a headland turn and refers to an "ideal" driver under soil conditions that do not allow for any slippage. The real traveled distance during a particular turn executed in the field (L) differs from the minimum by a stochastic amount ($\tilde{e}$): $L(d) = L_{min} + \tilde{e}$. The function $\tilde{e}$ incorporates the additional traveled distance due to factors such as driver dexterity, soil driving conditions, tire–ground surface interaction, vehicle dynamics, any activation of the braking system, etc. This function cannot be explicitly defined, and therefore is considered as a stochastic variable.

3.3.3   Modeling of Parallel Tracks Coverage System

The term "parallel field-work tracks system" refers to full coverage of a field by a set of parallel tracks of the same width (i.e., the working width of the machine) traversed by an agricultural vehicle, where each track starts from one end of the field (headland or maneuvering zone) and is completed at the opposite end. It is not necessary for the parallel paths to be straight sections; they may, for example, be curves parallel to the outer sides of the field. Also, the term "parallel" tracks is not strictly a geometric term (nonintersecting) because of the inevitable overlapped or missed areas occurring during a field operation and the potential presence of obstacles.

The parallel field-work tracks system should not be confused with the "row crop" system, where the term *system* refers to a seeding system deployed in rows, such as in cotton, sugar beet, etc., while in the former case the term *system* refers to the sequence of movements of the agricultural machine when working on the field.

The term "working width" (or "operating width"), which is denoted w_e, refers to the effective operating width and not the theoretical one, w_{th}. Letting u be the degree of exploitation of the working width, the effective operating width is given by $w_e = u \cdot w_{th}$. Depending on the operation and the plant and soil conditions, the effective operating width is normally between 90% and 95% of the theoretical one, and experienced drivers can reach the level of 95%–97%.[14] Navigation aids and autosteering systems can drastically increase the effective operating width.[49–51] In operations such as harvesting of row crops where there is a fixed number of cutting rows or in precision planting/seeding, there is no distinction between the effective and the theoretical operating widths. In fertilizer spreaders the operating width is the active width of fertilizer dispersion after any overlaps have been taken into account.

Let $T = \{1, 2, 3, \dots\}$ be the set of the field-work tracks which completely cover a particular field area. The counting direction is arbitrary. The number of tracks (the cardinality of set T) is given by:

$$|T| = \frac{\max(HL_u, HL_l)}{w_e} + 1$$

where HL_u, HL_l, are the lengths of the upper and lower headland area, respectively, and symbol $\lfloor \rfloor$ denotes the floor function. However, this number can be reduced by 1 either in the ideal case where the length of the headland area is an integer multiple of the operating width, or in a case where the farmer considers that the operating costs of an extra incomplete field-work track exceed the return from the yield produced by the particular track.

The mathematical form for the maneuvering degree, as defined in the previous paragraph, is:

$$d(i, j) = \lfloor i - j \rfloor, \; i, j \in T$$

3.3.4   Modeling Field Operations

Let us consider a field operation where the whole field area has to be worked by an agricultural machine with operating width w_e using the parallel field-work tracks system (Fig. 31). In this case a bijective function can be defined,[52] $p(\cdot) : T \rightarrow T$, which for each field-work track $i \in T$ returns the order in which the machine works the particular track. For example, in the case of a continuous pattern (Fig. 31A) the sequence function is identical (i.e., $p(i) = i$, $\forall i \in T; p(1) = 1; p(2) = 2$; etc.), while for particular alternate patterns the function values are $p(5) = 1; p(1) = 2, p(6) = 3$, etc.

The reverse function $p^{-1}(\cdot) : T \rightarrow T$ gives the field-work track sequence in which the agricultural vehicle covers the field area. For example, the statement $p^{-1}(3) = 6$ means that the third in order field-work track worked by the agricultural machine is the sixth field-work track on the field area. To this end, the sequence of the field-work tracks in a field operation is given by the enumeration $\sigma = <p^{-1}(1), p^{-1}(2), p^{-1}(3), \ldots p^{-1}(|T|)>$. For the pattern depicted in Fig. 31B the field-work track sequence is $\sigma = <5, 1, 6, 2, 7 \ldots>$, while in the continuous pattern shown in Fig. 31A the sequence is $\sigma = <1, 2, 3, \ldots>$.

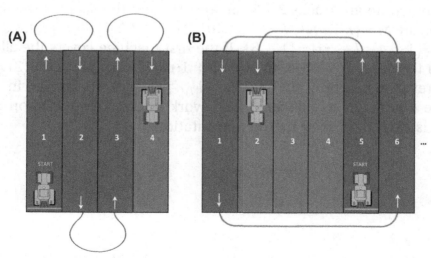

FIGURE 31 The area coverage system with parallel field-work tracks in continuous (A) and alternate (B) patterns.

3.3.5 Field Operations Executed by Multiple Machines

Consider that a field operation is executed by a number of agricultural machines of identical operational width. Let $M = \{1, ..., m\}$ denote the set of these machines. A set of field-work tracks, $T_k \subseteq T, k \in M$, corresponds to each machine k and defines the particular tracks that a machine has to work. For example, $T_3 = [4, 5, 6, 7]$ means the fourth, fifth, sixth, and seventh field-work tracks are allocated to the machine identified by index 3, while the number $|T_3| = 4$ gives the number of the allocated tracks. For full coverage of a field area (headlands excluded), the following conditions have to be satisfied:

$$\bigcup_M T_k = T$$

$$\bigcap_M T_k = \varnothing$$

By defining a general set $I_n = \{1, ..., n\}$, and analogous to the case of a single machine, for each machine $k \in M$ a bijective function can be defined, $p_k(\,\cdot\,) : T_k \rightarrow I_{|T_k|}$. For each field-work track $i \in T_k$ that has been allocated to machine k, the number $p_k(i)$ represents the order in which the ith track of the field is worked by this machine. For example, for two machines m_1 and m_2 operating in a field of 10 field-work tracks, the tracks allocated to the first machine are 1,3,5, and 7 and those allocated to the second machine are tracks 2,4,6,8,9, and 10; and the sets for the specific operation are $M = \{m_1, m_2\}$, $T = \{1, ..., 10\}$, $T_1 = \{1, 3, 5, 7\}$, $T_2 = \{2, 4, 6, 8, 9, 10\}$, $I_{|T_1|} = I_4 = \{1, ... 4\}$, $I_{|T_2|} = I_6 = \{1, ... 6\}$. If the first machine covers the allocated tracks in the order 1−5−7−3, then $p_1(1) = 1; p_1(5) = 2; p_1(7) = 3; p_1(3) = 4$.

The reverse function $p_k^{-1}(\,\cdot\,) : T_k \rightarrow I_{|T_k|}$ provides the order in which machine k operates its allocated field-work tracks. The operation of machine k is then described by the permutation:

$$\sigma_k = < p_k^{-1}(1), p_k^{-1}(2), \,..., p_k^{-1}(|T_k|) >$$

3.3.6   In-Field Nonworking Traveled Distance

From the previous definitions of the turning degree and the parallel tracks coverage function $p(\cdot): T \to T$, one can derive that the turning degree can be expressed as:

$$d(i, i+1) = \left| p^{-1}(i+1) - p^{-1}(i) \right|, \ i, j \in T$$

Based on this, for a permutation σ of a set of field-work tracks for complete area coverage, the total distance that the machine travels during the headland turnings is given by:

$$J(\sigma) = \sum_{i-1}^{|T|-1} L_{min}\left(\left| p^{-1}(i+1) - p^{-1}(i) \right| \right) = \sum_{i-1}^{|T|-1} L_{min}(d(i, i+1))$$

For multiple machines, the previous function is written as:

$$J(\sigma_1, \ldots, \sigma_m) = \sum_{k=1}^{m} \sum_{i-1}^{|T_k|-1} L_{min}\left(\left| p_k^{-1}(i+1) - p_k^{-1}(i) \right| \right) = \sum_{k=1}^{m} \sum_{i-1}^{|T_k|-1} L_{min}(d_k(i, i+1))$$

This function requires one part of the objective function to be minimized in optimized route planning. It represents the total in-field nonworking (or noneffective) traveled distance. The second part of the objective function corresponds to the distances the machine has to travel to move from its initial position (i.e., field entrance) to the position where the operation commences, and from the position where the operation is completed to the final position (i.e., field exit point).

3.3.7   Basic Routing Optimization Problem

3.3.7.1   Algorithms

The problem of finding the optimal tracks sequence of the basic problem, as defined in the previous Section, is identical to the well-known combinatorial optimization problem, the travelling salesman problem (TSP). For all the extended problems (e.g., multiple machines, orchard operations, etc.) the basic problem constitutes a subcase. Consequently, all agricultural machinery area coverage problems in the parallel field-work tracks system present at least the complexity of the TSP and belong to the NP-hard[2] class

---

[2]"NP-hard" stands for "non-deterministic polynomial-time hard," and is a term used in computational complexity theory.

of combinatorial optimization problems. To depict the complexity of this type of problem, if we consider a problem with n nodes (or field-work tracks), the number of combinations of the feasible closed-loop routes that constitute the solution space of the problem is $(n-1)!/2$. An algorithm that could potentially evaluate exhaustively all feasible solutions of the problem to select the optimal one would require approximately $n!$ computational steps. This means for a field area coverage problem of a field dimension/machine operating width combination with 20 field-work tracks, 20! (2,432,902,008,176,640,000) computational steps are required, while in the case of 50 field-work tracks the number climbs up to 50!, a number with 65 digits (3.041×10^{64}).

To address this category of combinatorial optimization problems, algorithms have been developed based on various methods to produce solutions within a reasonable time. The modeling of the field area coverage problem enables the use of graph search (traversal) algorithms, and also integer algorithms for solving integer programming optimization problems. The algorithms used can be divided into two general categories based on the balance between the computational time and the quality, in terms of optimality, of the generated solution.

In the various cases for generating an optimal solution to a field-work planning problem, there are different requirements for the quality of the solution and the speed of its generation. For example, for a plowing operation in a particular field using a particular machine, the plan can be produced at any time before the actual operation, thus a fast-generated solution is not a prerequisite. On the other hand, if planning takes place immediately after on-site identification of a given parameter (e.g., soil moisture or the presence of obstacles), low computational time is required. Consequently, in the case of online (or real-time) planning, priority is given to the speed of the solution generation at the expense of the optimality of the solution.

The algorithms generally used to solve problems of combinatorial optimization can be divided into two basic categories[53]: exact and approximate (or heuristic). Exact algorithms always give the best solution in a limited number of steps. These algorithms are generally quite complex, and place increased demands on computing resources (time and memory).[54-56]

Approximate algorithms are generally subcategorized into three classes:

- tour construction algorithms
- tour improvement algorithms
- composite algorithms.

Algorithms of the first class build a route in a step-wise manner by continuously adding new nodes.[57] Algorithms of the second class improve a route by implementing various interchanges of the node sequence.[58] Finally, algorithms of the third class combine the methodological features of algorithms of the other two classes.

Descriptively, the three basic steps of a randomized search in the problem solution process are as follows.

- Initializing: create and evaluate an initial set of feasible possible S solutions.
- Function: generate and evaluate a new set of feasible solutions, S', through randomized changes to selected members in S.
- Refresh: replacement of some members in S by some members of S', and return to the second step or finish the process.

The branch-and-bound method for the solution of an integer binary programming problem belongs to the category of exact algorithms. It seeks the best solution to a binary integer programming optimization problem by solving a number of linear problems (LP relaxation), where the "hard" constraint for the decision variables to be of values of either 0 or 1 is replaced with the "softer" constraint of being real numbers between 0 and 1. The process of finding a solution can be summarized as follows:

- the algorithm looks for a feasible solution for the binary problem
- it continuously updates the best possible solution point as the search tree grows
- it ensures that there is no better possible integer solution by solving a series of linear programming problems.

More analytically, each of these three steps includes two stages.[59,60]

Branching. The algorithm creates a search tree as it iteratively adds new boundaries to the problem. In each branching step the algorithm selects a variable, x_i, and adds the restriction $x_i = 0$ to create one branch and the restriction $x_i = 1$ to create another branch. This process can be illustrated by a binary tree in which the nodes represent the added restrictions. For example, the tree shown in Fig. 32 represents a full binary tree for a problem that has three variables, x_1, x_2, and x_3. At each node the algorithm solves a linear programming problem (LP relaxation) using the constraints of this node and decides, depending on the result, either to continue on this branch or move to another node. If the LP relaxation problem at that particular node has no feasible solution or the objective value of the solution is greater than the best objective value of an integer decision variable, the algorithm removes the node from the tree and does not search for any branches following this node in the search tree. If a new feasible integer point is found with a lower objective value from the best integer point, the algorithm updates the best solution and proceeds to the next node.

Bounding. The solution of the LP relaxation problem gives a lower bound to the problem of binary integer programming. If the solution is an integer vector, it gives an upper bound for the binary integer programming problem. As the tree search proceeds, the algorithm updates the lower and upper bounds of the objective function by using the updated new boundaries resulting from the bounding step, which act as a threshold for cutting out the branches that lead away from the best solution.

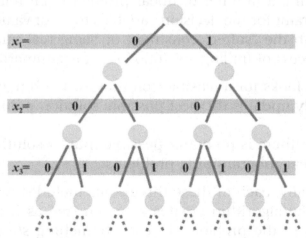

FIGURE 32 Search tree generation.

3.3.7.1.1   CLARKE–WRIGHT ALGORITHM (MODIFIED)

The particular algorithm presented here is a modification of the well-known Clarke–Wright algorithm.[61,62] The algorithm was first developed to address the problem of optimum route generation for a fleet of trucks, which are heterogeneous in terms of carrying capacities, from a central depot to a number of delivery points. The algorithm is based on an iterative procedure that enables rapid selection of an optimum or near-optimum route. The algorithm operates in two stages.

Randomization: In the first stage, starting from the first node and for each step of the path in the graph, the most economical (least cost) connection to the next node is selected. The classic algorithm can be extended if at each step it randomly chooses one of k best connections and the process is repeated several times before choosing the best solution. More specifically, in the first step a number of iterations i and a number j corresponding to the depth of the iterations are defined. In the first iteration of the most economical connection is selected at each step. In the second iteration one of the two most economical connections are randomly chosen at each step, and i solutions are generated in this way. During the third iteration, one of the three most economical connections is randomly selected at each step, and i solutions are generated. This continues until solutions have been generated by random selections between the j most economical combinations. In total $i \times j$ solutions are generated, but are not necessarily nonidentical. A case where $i = j = 1$ corresponds to the classic Clarke–Wright algorithm.

Heuristics: After creating the first solution from the randomization stage, various heuristic operations are applied.

- R-opt. This is one of the best-known heuristic manipulations, where r arcs are selected in the solution path, removed from the particular path, and replaced by r other arcs. Fig. 33 presents an example for the case of R = 2 (2-opt).
- Or-opt. These operations use groups of nodes of populations: $l \in \{1, 2, 3\}$.[63] In these operations, part of the route consisting of l subsequent nodes is removed and inserted in another part of the route. Fig. 34 gives an example of Or-opt operation. The square in the figure represents the warehouse; the left part of the figure depicts the route before the Or-opt operation implementation, while the right

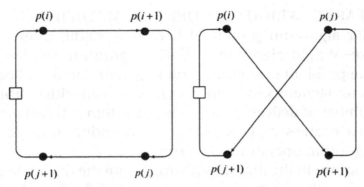

FIGURE 33 An example of a 2-opt operation.

part depicts the route after implementation. The sequencing of the nodes from $p(i+1)$ to $p(i+l)$ remains the same.

- Swap operations. This operation applies to routing problems with capacity constraints or using multiple vehicles. During the swap operation, two nodes from different routes are exchanged by replacing one with the other.[64] Fig. 35 gives an example of a swap operation in a problem involving two routes. The squares in the figure represent the depot; the left figure shows the solution before the implementation of the swap operation, while on the right is the solution (new route) after implementation of the swap. The number of nodes between $p_1(i)$ and $p_2(j)$ can by any, as well as between the nodes $p_1(1+i)$ and $p_2(1+j)$.

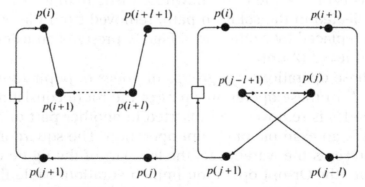

FIGURE 34 An example of Or-opt operation.

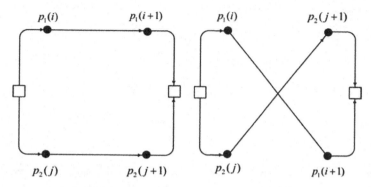

FIGURE 35 An example of swap operation.

3.3.7.2   Field Shape

The fields presented in Fig. 36 are abstractions of typical field shapes which have been used in the agricultural operations management literature for various simulated experiments.[3,65,66] The fields in the figure are all of 10 ha area.

To quantify the performance of machinery, two indices are defined: distanced-based field efficiency (DFE) as a function of the field shape, machinery features, operating width, and fieldwork pattern, and field coverage ratio (AI).

Specifically, $DFE = D_{effective}/D$, where the total effect is $D_{effective} = D - D_{non-working}$.

However, the effective traveled distance might lead to some overlap area; for example, the last track might not generate a whole row within the field body area.[67] To take this into account, the field coverage ratio index is given by:

$$AFE = \frac{field\ area}{D_{effective} \times w}$$

Table 16 presents the values of the DFE and area-based field efficiency (AFE) for the same fields and machinery features as in Table 15.

3.4 Machinery Systems' Productivity

In a system consisting of a series of subsystems, if one of the component subsystems fails the whole system fails. This is the reason why in a

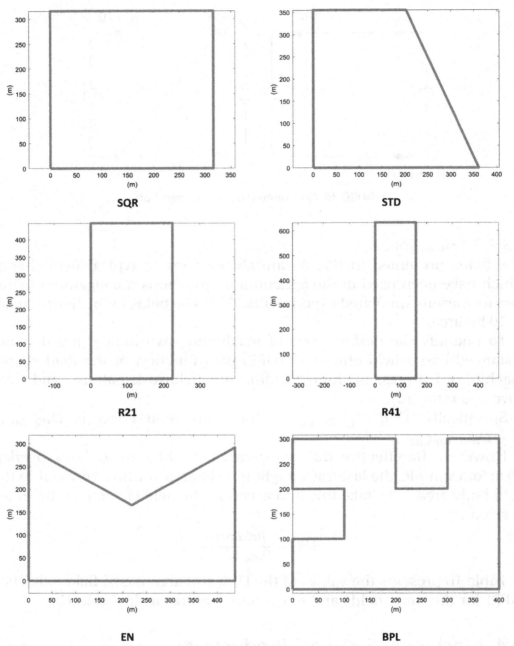

FIGURE 36 Templates representing typical fields. SQR: square; STD: standard; R21: rectangular 2:1; R41: rectangular 4:1, REN: reentrant; BPL: building plot.

Table 15 Productivity and Area Capacity Variations for Template Field Shapes (10 ha)[3]

Shape	Productivity (min ha^{-1})	Productivity Index	Area Capacity (ha h^{-1})	Area Capacity Index
SQR	56.6	100	1.06	100
R21	54	95	1.11	105
R41	52.4	93	1.15	103
STD	59.5	105	1.01	88
REN	59.1	104	1.02	101
BPL	60.5	107	0.99	98

Table 16 Distance-Based Field Efficiency and Area-Based Field Efficiency for Template Fields and Corresponding Indexes for a Machine of 3 m Working Width and 5 m Turning Radius

Shape	Total Distance (m)	Total Effective Distance (m)	Total Nonworking Distance (m)	DFE	DFE Index	Overlapped Area (m^2)	AFE Index
SQR	36955	33168	3788	0.90	100		0.90
R21	36388	33483	2905	0.92	103	448.6	0.92
R41	35730	33467	2263	0.94	102	400.5	0.93
STD	37583	33359	4224	0.89	95	76.7	0.89
REN	38589	33397	5192	0.87	98	190.7	0.86
BPL	40021	33353	6668	0.83	96	58.1	0.83

machinery system the productivity should be considered for the system as a whole and not for any individual machine performing part of the operation. To this end, an analysis of the system has to be made. Various tools have been developed for the operational description of systems that includes elements operating both in parallel and series manner. One of the most common tools applied in complex systems is IDEF; this was initially the abbreviation for ICAM Definition (where ICAM is an acronym for integrated computer-aided manufacturing) and was renamed "Integration Definition" in 1999. IDEF includes a number of definition languages, IDEF0, IDEF2, IDEF3, IDEF4, and IDEF5, which were developed with US Air Force funding and are widely implemented in military operations research.[68] IDEF has been implemented to describe various agricultural operations.[69,70]

There two types of IDEF schematics, namely the IDEF process schematic that displays a process-centered view of a system (or organization) and the IDEF object schematic that displays an object-centered view of the system (or organization). The former has been used for schematic descriptions of various agricultural operational systems. IDEF process schematics capture, manage, and display knowledge about events and activities based on the various processes taking place within the operational system, all objects that are related to these processes, and any constraining relations governing the system behavior.

Fig. 37 presents the elements of a building block in an IDEF0 semantic description of a system. All functions of the system are governed by input, mechanism, control, and output elements, and are connected

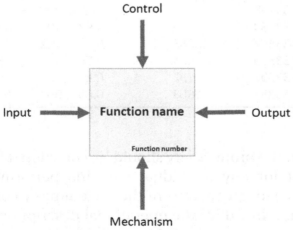

FIGURE 37 Building block of IDEF0.

by junctions. Table 17 lists three junctions that are widely used for the semantic representation of agricultural operations executed by a machinery system.

Table 17   Selected[3] Junctions Implemented in IDEF3

Icon	Junction Name	Fan-In (Convergence Flow)	Fan-Out (Divergence Flow)
&	Asynchronous AND	All the preceding parallel activities must be completed.	All the following parallel activities must begin
O	Asynchronous OR	One or more of the preceding alternative activities must be completed	One or more of the following alternative activities must begin
X	Exclusive OR (XOR)	Exactly one preceding activity of mutually exclusive alternatives must be completed	Exactly one following activity begins

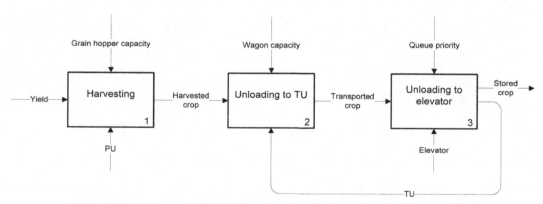

FIGURE 38 A general IDEF0 diagram for the crop harvesting process.

[3]Beyond the presented junctions, the synchronous AND and synchronous OR are also used. For an in-depth study of IDEF3 see Mayer et al..[68]

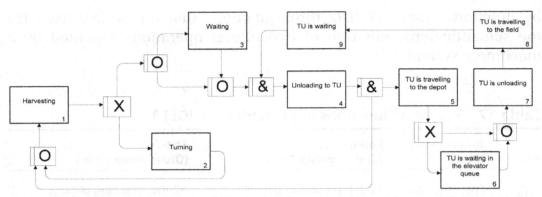

FIGURE 39 An IDEF3 description of the harvesting operation with out-of-field stationary unloading.

As examples of IDEF implementation in agricultural operations, two cases are presented: a simplified description of the harvesting operation (Fig. 38) and a more complex description of the harvesting operation with a stationary out-of-field unloading process (Fig. 39).

4

Cost of Using Agricultural Machinery

4.1 Direct Cost

4.1.1 Ownership Cost

4.1.1.1 Depreciation

Depreciation is the reduction of the value of an asset due to time, usage, and/or obsolescence (technological outdating). There are many methods to calculate depreciation.

4.1.1.1.1 STRAIGHT-LINE METHOD

In the straight-line method, the total depreciation of an asset unit is uniformly distributed along its useful life. Based on this assumption, the annual depreciation is given by:

$$D(i) = D = \frac{Q_0 - Q_s}{n} \quad \forall\, i \in \{1....n\}$$

where $D(i)$ is the annual depreciation for the year i (for the particular method the depreciation value is identical for any year of the useful life), Q_o is the purchase value of the asset unit, Q_s is the salvage value[a] of the unit at year i, and n is the period of ownership (useful life of the asset).

The straight-line method is the simplest way to estimate the depreciation cost. However, the estimated value of the asset during the initial years of the ownership (or economic life) is overestimated compared to the real value of the asset in the second-hand market.

[a]Also known as "trade-in value" and "scrap value". Usually, if not otherwise given. $Q_s = 0.1 \cdot Q_0$

Operations Management in Agriculture. https://doi.org/10.1016/B978-0-12-809786-1.00004-7

The book value of the asset at year i is given by:

$$Q_n(i) = Q_0 - \frac{Q_0 - Q_s}{n} \cdot i = Q_0 - i \cdot D$$

It must be mentioned that the purchase value of the asset is the purchase price, whether it is a new or a second-hand machine. If the asset is sold before the end of its economic life, the resale value is its residual value. The resale value is now the purchase value for the new owner.

Example: Consider a machine of €10000 purchase value, useful life $n = 10$ years, and remaining value of $Q_s = 0.1 \cdot Q_0$. Based on the straight-line depreciation method, for the whole useful life of the machine the annual depreciation will be:

$$D = \frac{10000 - 0.1 \cdot 10000}{10} = 900 \text{ €}$$

At the end of the fourth year ($i = 4$) the book value of the machine is estimated to be:

$$Q_{10}(4) = 10000 - 4 \cdot 900 = 6400 \text{ €}$$

Note that the straight-line method, due to the fixed annual depreciation, overestimates the remaining value of the machine during the first years of its economic life.

4.1.1.1.2 SUM-OF-YEARS DIGITS METHOD

This is one of the accelerating depreciation methods, and is based on the general assumption that the productivity of an asset is higher during the first period of its productive life since it is new.

The summation of the digits of the useful life (n) of the asset is given by:

$$SYD = \sum_{i=1}^{n} i = \frac{n^2 + n}{2}$$

The depreciation is given by:

$$D(i) = (Q_0 - Q_s) \cdot \frac{(n - i + 1)}{SYD} \quad \forall i \in \{1....n\}$$

$n - i$ represents the remaining useful life of the asset.

The book value of the asset at year i is given by:

$$Q_n(i) = Q_0 - (Q_0 - Q_s) \cdot \frac{i \cdot (2 \cdot n - i + 1)}{n \cdot (n + 1)}$$

Example: Using the same input as in the example presented above, the summation of the digits of the useful life of the machine is $SYD = 55$. Based on that, the depreciation for the first 4 years is:

$$D(1) = (10000 - 1000) \cdot \frac{(10 - 1 + 1)}{55} = 1636.4 \text{ €}$$

$$D(2) = (10000 - 1000) \cdot \frac{(10 - 2 + 1)}{55} = 1472.7 \text{ €}$$

$$D(3) = (10000 - 1000) \cdot \frac{(10 - 3 + 1)}{55} = 1309.1 \text{ €}$$

$$D(4) = (10000 - 1000) \cdot \frac{(10 - 4 + 1)}{55} = 1145.5 \text{ €}$$

The book value at the end of the fourth year is:

$$Q_{10}(4) = 10000 - (10000 - 1000) \cdot \frac{4 \cdot (2 \cdot 10 - 4 + 1)}{10 \cdot (10 + 1)} = 4436.4 \text{ €}$$

The book value with this method is considerably lower than the corresponding value found by using the straight-line method for depreciation estimation, yet more realistic for the first years of the machine ownership.

4.1.1.1.3 SINKING FUND METHOD

The sinking fund method is based on the assumption that an amount corresponding to the yearly depreciation is deposited in a bank account and earns compound interest every year. In this way, the sum of the initial capital invested to purchase the asset unit, the interest, and the remaining value corresponds to the initial value of the unit (recovering invested initial capital). The sinking fund method is the most rational method, but it requires an account into which regular deposits are made—something that is rarely observed.

According to the sinking fund method, the annual depreciation is given by:

$$D = \frac{(Q_0 - Q_s) \cdot \varepsilon}{(1 + \varepsilon)^n - 1}$$

where ε the annual interest rate.

The book value of the asset at year i is given by:

$$Q_n(i) = Q_0 - (Q_0 - Q_s) \cdot \frac{(1+\varepsilon)^i - 1}{(1+\varepsilon)^n - 1}$$

Example: Using the input of the examples presented above, the annual interest rate is $\varepsilon = 4\%$. Based on that, the annual depreciation is:

$$D = \frac{(10000 - 1000) \cdot 0.04}{(1 + 0.04)^{10} - 1} = 750 \; €$$

and the book value at the end of the fourth year is:

$$Q_{10}(4) = 10000 - (10000 - 1000) \cdot \frac{(1 + 0.04)^4 - 1}{(1 + 0.04)^{10} - 1} = 6815.1 \; €$$

4.1.1.1.4 DECLINING BALANCE OR FIXED PERCENTAGE METHOD

With the declining balance method, the annual depreciation represents a fixed percentage of the book value of the machine over time. With this method the reduction in the machine value is faster in the first years of the asset's economic life. Over the years the depreciation value continuously decreases.

The percentage of annual depreciation q is:

$$q = 1 - \sqrt[n]{\frac{Q_s}{Q_o}}$$

where Q_o is the purchase value of the asset unit, Q_s is the salvage value of the unit at year i, and n is the period of ownership (useful life of the asset).

The annual depreciation $D(i)$ for the year i is given by the equation:

$$D(i) = Q_o \cdot \left(1 - \sqrt[n]{\frac{Q_s}{Q_o}}\right) \qquad \forall i \in \{1....n\}$$

At the end of the ith year the book value is given by the equation:

$$Q_n(i) = Q_0 \cdot \left[1 - \left(1 - \sqrt[n]{\frac{Q_s}{Q_o}}\right)\right]^t$$

Example: Using the input of the examples presented above, the depreciation at the end of the fourth year is:

$$D(4) = 10000 \cdot \left(1 - \sqrt[10]{\frac{1000}{10000}}\right) = 2057 \text{ €}$$

The book value at the end of the fourth year is:

$$Q_{10}(4) = 10000 \cdot \left[1 - \left(1 - \sqrt[10]{\frac{1000}{10000}}\right)\right]^4 = 3980.5 \text{ €}$$

American handbooks mention a similar method called the double-declining balance. With this method there is fast depreciation during the first years of the machine's economic life.

4.1.1.1.5   RESALE VALUE OR ESTIMATED VALUE METHOD

The resale value method is the most realistic. According to this method, at the end of each year the value of the asset is compared with the value at the beginning of the year. The difference constitutes the reduction of its value (the annual depreciation). For this method to be reliable, a realistic estimate of the value of the machine should be made. The best way is to assess the value based on the resale value of the machinery in the resale market. It is well known that the resale value is affected by use and the repairs/maintenance of the machinery.

$$RV_n = 100 \cdot \left[C_1 - C_2 \cdot (n^{0.5}) - C_3 \cdot (h^{0.5})\right]^2$$

where RV_n is the remaining value (resale value) (%), n is the end of the year, h is the average operating hours per year, and C_1, C_2, C_3 are the remaining value coefficients.

Table 18 gives the remaining value coefficients for different machinery.

The remaining value of the machinery as a percentage of the purchase value can be estimated from the ASAE Standards[71] equations (Table 19).

4.1.1.2   *Inflation*

In periods with high annual inflation rates, this inflation should be taken into account to calculate the asset value. It is necessary to refer to stable values related the year of purchase or to corrected values at the year of calculation.

Table 18 Remaining Value Coefficients[44]

Equipment Type	C1	C2	C3
Farm Tractors			
Small <60 kW (80 hp)	0.981	0.093	0.0058
Medium 60–112 kW (80–150 hp)	0.942	0.100	0.0008
Large >112 kW (150 hp)	0.976	0.119	0.0019
Harvesting Equipment			
Combines	1.132	0.165	0.0079
Mowers	0.756	0.067	—
Balers	0.852	0.101	—
Swathers and all other harvesting equipment	0.791	0.091	—
Tillage Equipment			
Plows	0.738	0.051	—
Disks and all other tillage equipment	0.891	0.110	—
Miscellaneous Equipment			
Skid-steer loaders and all other vehicles	0.786	0.063	0.0033
Planters	0.883	0.078	—
Manure spreaders and all other miscellaneous equipment	0.943	0.111	—

Table 19 Remaining Values as Percentage of List Price at End of Year n[72]

Equipment Type	
Tractors	$68 \cdot (0.920)^n$
Combines, cotton pickers, self-propelled (SP) windrowers	$64 \cdot (0.885)^n$
Balers, forage harvesters, blowers, and SP sprayers	$56 \cdot (0.885)^n$
All other field machines	$60 \cdot (0.885)^n$

The inflation coefficient for n years is:

$$Inf_{coef}(n) = \prod_{i=1}^{n}[1 + ir(i)]$$

where $ir(i).i \in \{1.....n\}$ is the yearly inflation rate.

If the inflation rate is taken into account, the purchase value of the asset unit has to be corrected accordingly:

$$Q = Inf_{coef} \cdot Q_0$$

Example: Following the same example, if the straight-line method for depreciation and assuming for the 3 years that the inflation rates are $ir(1) = 6\%$, $ir(2) = 4\%$, $ir(3) = 3\%$, the inflation coefficient amounts to $Inf_{coef} = (1 + 0.06) \cdot (1 + 0.04) \cdot (1 + 0.03) = 1.135$.

The corrected initial price of the machine is $Q = 10000 \cdot 1.135 = 11350$ €. Based on this, the annual depreciation of the machine equals:

$$D = \frac{0.9 \cdot 10000}{10} = 900 \text{ €}$$

Moreover, inflation also affects the real interest, which is written as:

$$\varepsilon_r(i) = \frac{\varepsilon_n - ir(i)}{1 + ir(i)}$$

where ε_r is the real interest rate and ε_n is the investment interest rate.

Example: Considering the annual inflation rates of the previous example, i.e., $ir(1) = 6\%$, $ir(2) = 4\%$, $ir(3) = 3\%$, and an investment interest rate of 8% which is stable throughout the 3-year period, the annual real interest rates corresponding to the 3 years are:

$$\varepsilon_r(1) = \frac{0.08 - 0.06}{1 + 0.06} = 1.89\%$$

$$\varepsilon_r(2) = \frac{0.08 - 0.04}{1 + 0.04} = 3.85\%$$

$$\varepsilon_r(3) = \frac{0.08 - 0.03}{1 + 0.03} = 4.85\%$$

4.1.1.2.1 COMPARISON OF METHODS

Comparison of the depreciation methods shows that the declining balance method and sum-of-years digits method give similar results. In both methods the depreciation is very high in the early years to approach the value in the second-hand market. The sinking fund method gives low depreciation in early years and higher depreciation in later years. Finally, the straight-line method gives steady depreciation throughout the economic life.

Table 20 Comparison Between Various Depreciation Methods

Year	Straight-Line Method	Sum of Digits Method	Sinking Fund Method	Declining Balance Method	Resale Value Method
1	0.91	0.84	0.93	0.79	0.53
2	0.82	0.69	0.85	0.63	0.47
3	0.73	0.56	0.77	0.50	0.42
4	0.64	0.44	0.68	0.40	0.37
5	0.55	0.35	0.59	0.32	0.33
6	0.46	0.26	0.50	0.25	0.29
7	0.37	0.20	0.41	0.20	0.26
8	0.28	0.15	0.31	0.15	0.23
9	0.19	0.12	0.21	0.12	0.20
10	0.10	0.10	0.10	0.10	0.18

Table 20 and Fig. 40 present the remaining values of an agricultural machine using all the depreciation methods. The purchase value (Q_o) is 100 units, the salvage value (Q_s) is 10% of the initial price, the economic life is 10 years, and the interest is 4%.

4.1.1.3 Interest

To purchase an asset unit, the capital (or part of it) is either borrowed or owned. In the former case interest on the investment has to be paid, while

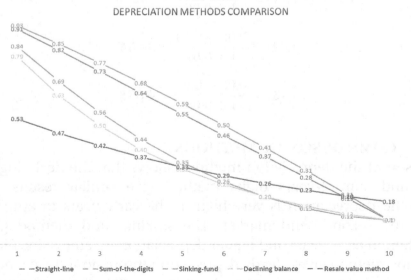

FIGURE 40 Remaining values of an agricultural machine using all the depreciation methods for an economic life of 15 years.

in the latter case there is an opportunity cost of the investment representing the lost return that the capital could earn from an alternative investment. Although measuring the opportunity cost is more or less hypothetical, a realistic and practical approach is to consider as opportunity cost the earning power of a safe investment such as a bank deposit. To this end, when estimating the ownership cost of an asset unit the interest charge has to be taken into account (Fig. 41).

The interest charge (*IC*) for a year is obtained from the average investment on the current year, which is half the sum of the book value at the beginning and the end of the year multiplied by the real interest rate if the inflation rate is taken into account:

$$IC = \frac{Q_i + Q_{i-1}}{2} \cdot \varepsilon_r$$

where Q_i the asset book value on year i and ε_r is the real interest rate (decimal).

Example: Ownership cost estimation based on the following facts for an asset unit:

- purchase value, $Q_0 = 50000 €$
- average (for the first 10 years) yearly inflation rate of 3%, $ir(i) = 0.03$, i $= \{1.....10\}$

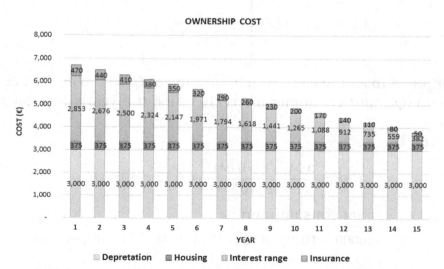

FIGURE 41 Yearly ownership cost for all the useful life.

- useful life, n = 15 years
- salvage value, 10% of the initial price, $Q_s = 0.1 \cdot Q_0$

$$IC = \frac{Q_i + Q_{i-1}}{2} \cdot \varepsilon_r$$

4.1.1.4 Capital Consumption Method

A method for determining the capital cost of ownership is based on the capital recovery factor, *CRF*, which represents the total depreciation of the asset unit and the capital interest at the end of its useful life. The CRF is given by:

$$CRF = \frac{\varepsilon \cdot (1 + \varepsilon)^n}{(1 + \varepsilon)^n + 1}$$

when considering equal yearly installments. In this case the yearly capital consumption is given by:

$$CC = (Q_0 - Q_s) \cdot CRF + Q_s \cdot \varepsilon$$

In a case where a series (*q*) of equal payments over the life of the asset is considered, the capital consumption is given by:

$$CC = (Q_0 - Q_s) \cdot \left[\frac{\frac{\varepsilon}{q}}{1 - \left(1 + \frac{\varepsilon}{q}\right)^{-nq}} \right] + Q_s \cdot \frac{\varepsilon}{q}$$

Example: A machine has a purchase value of $Q_0 = 10000$, salvage value of $Q_s = 0.1 \cdot Q_0$, and economic life of 10 years; the yearly interest rate is $\varepsilon = 4\%$.

The CRF is:

$$CRF = \frac{0.04 \cdot (1 + 0.04)^{10}}{(1 + 0.04)^{10} - 1} = 0.12 = 12\% \, Q_0$$

The yearly capital consumption is:

$$CC = (10000 - 1000) \cdot 0.12 + 1000 \cdot 0.04 = 1120 \, \euro \text{ per year}$$

The value is the machine occupation annual cost, paid in equal installments every year until the end of the economic life of the machine.

4.1.1.5 Housing

All agricultural machinery needs shelter: this slows wear of the machines, their repair and maintenance are easier, and their resale price is increased. With today's high machine prices, the construction of a new outbuilding is not a luxury.

Machinery can be housed in fairly simple constructions, so the costs are similar; they include depreciation of the invested capital, maintenance cost, and the interest on the circulating capital.

For the calculation of depreciation and maintenance costs, various subscales must be taken into account.

- Wooden outbuilding: 10-year lifetime and maintenance costs 5% of the initial price.
- Metal outbuilding: 20-year lifetime and maintenance costs 3% of the initial price.
- Outbuilding with reinforced concrete: 40-year lifetime with maintenance costs 0.5% of the initial price.

The dimensions of the outbuilding depend on the size of the machinery. For large tractors the height should be more than 4 m, length 5–6 m, and width 3.5–4 m.

If there is no accounting data, housing maintenance costs can be estimated at 0.5%–1% of the machine purchase price.[14] If an old building is used for housing, its maintenance costs can be assessed as 0.2%–0.5% of the machine purchase price. If there is a particular room for the repair of machinery and storage of fuel, lubricants, and spare parts, the cost of this space should be added to the cost of the housing.

4.1.1.6 Insurance and Taxes

Agricultural machines, especially self-propelled machines such as tractors and harvesters, and their operators are constantly exposed to the risks of overturning, fire, collisions, and falling into irrigation or drainage networks, resulting in machine damage, operator injuries, and third-party damages. It is thus necessary to insure them against these risks. If they are not insured, the cost analysis should nonetheless include the insurance cost, since the owner is the person who pays the cost in any case of damage.

Generally, self-propelled agricultural machinery should have full insurance covering third-party injury or damage, damage to the machine, and operator and/or codriver injury.

The insurance cost is 0.5%–1% of a machine's initial purchase price. If the average value is about 50% of the initial price, an average charge of 0.25%–0.5% of the initial price can be considered reasonable.

4.1.2 Operating Cost

The operating or variable cost refers to the operation of the machine and occurs only when the machine is used. The term *operating cost* is usually used to describe the variable cost. The variable cost is usually calculated on an hourly basis, but may refer to processed area (e.g. ha), bales, or any other appropriate unit.

4.1.2.1 Repair and Maintenance Cost

Repair and maintenance are related to machine reliability: that it will carry out a cycle duty without any total or partial failure. Repair and maintenance functions can be categorized as follows[10,31]:

- routine replacement of worn parts
- repair of accidental damage
- repair of damage due to operator neglect
- routine overhauls.

The accumulated repair and maintenance cost (RM_{ac}) of an agricultural machine is given by the equation:

$$RM_{ac} = Q_o \cdot RF1 \cdot U_{ac}^{RF2}$$

where $RF1$ is the repair constant factor, $RF2$ is the repair exponent factor, and U_{ac} is the accumulated use of the machine. Repair factors for various types of agricultural machines are listed in Table 21.

The annual repair and maintenance cost is not stable throughout the useful life of the machine, but usually follows an exponential trend with lower yearly cost in the first years of use and a higher cost at the end of the useful life (Figs. 42–44).

Example: A two-wheel drive (2WD) tractor with a purchase value of €50000; the tractor is at the end of its seventh year and is used for 500 h

Table 21   Repair and Maintenance Cost Parameters[72]

Machine Tractors	Estimated Life (h)	RF1	RF2	Total Life Repair and Maintenance Cost (Q_o Ratio)
2WD tractor	12000	0.007	2.0	1.01
4WD tractor	16000	0.003	2.0	0.77
Tillage and Planting				
Moldboard plow	2000	0.29	1.8	1.01
Heavy-duty disk	2000	0.18	1.7	0.58
Tandem disk harrow	2000	0.18	1.7	0.58
(Coulter) chisel plow	2000	0.28	1.4	0.74
Field cultivator	2000	0.27	1.4	0.71
Spring-tooth harrow	2000	0.27	1.4	0.71
Roller—packer	2000	0.16	1.3	0.39
Mulcher—packer	2000	0.16	1.3	0.39
Rotary hoe	2000	0.23	1.4	0.61
Row crop cultivator	2000	0.17	2.2	0.78
Rotary tiller	1500	0.36	2.0	0.81
Row crop planter	1500	0.32	2.1	0.75
Grain drill	1500	0.32	2.1	0.75
Harvesting				
Corn picker—sheller	2000	0.14	2.3	0.69
Combine	2000	0.12	2.3	0.59
Combine (self-propelled)	3000	0.04	2.1	0.40
Mower	2000	0.46	1.7	1.49
Mower (rotary)	2000	0.44	2.0	1.76
Mower—conditioner	2500	0.18	1.6	0.78
Mower—conditioner (rotary)	2500	0.16	2.0	1.00
Windrower (self-propelled)	3000	0.06	2.0	0.54
Side-delivery rake	2500	0.17	1.4	0.61
Rectangular baler	2000	0.23	1.8	0.80
Large rectangular baler	3000	0.10	1.8	0.72
Large round baler	500	0.43	1.8	0.12
Forage harvester	500	0.15	1.6	0.05
Forage harvester (self-propelled)	4000	0.03	2.0	0.48
Sugar beet harvester	1500	0.59	1.3	1.00
Potato harvester	500	0.19	1.4	0.07
Cotton picker (self-propelled)	3000	0.11	1.8	0.79

Continued

Table 21   Repair and Maintenance Cost Parameters[72]—cont'd

Machine	Estimated Life (h)	RF1	RF2	Total Life Repair and Maintenance Cost (Q_o Ratio)
Miscellaneous				
Fertilizer spreader	1200	0.63	1.3	0.80
Boom-type sprayer	1500	0.41	1.3	0.69
Air-carrier sprayer	2000	0.20	1.6	0.61
Bean puller—windrower	2000	0.20	1.6	0.61
Beet topper/stalk chopper	1200	0.28	1.4	0.36
Forage blower	500	0.22	1.8	0.06
Forage wagon	2000	0.16	1.6	0.49
Wagon	3000	0.19	1.3	0.79

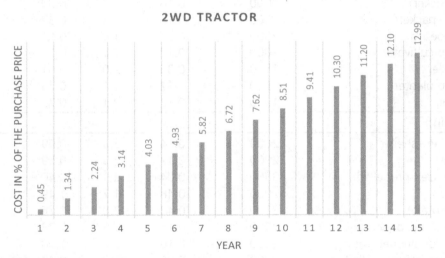

FIGURE 42 Yearly repair and maintenance cost expressed as a percentage of the purchase price for a 2WD tractor with useful life of 15 y and annual use of 800 h y^{-1}.

per year. The repair and maintenance cost is calculated in the following equations.

The cumulative cost for the first 7 years is:

$$RM_{ac} = Q_o \cdot RF1 \cdot U_{ac}^{RF2} = 50000 \cdot 0.007 \cdot \left(\frac{7 \cdot 500}{1000}\right)^{2.0} = 4287.5 \text{ €}$$

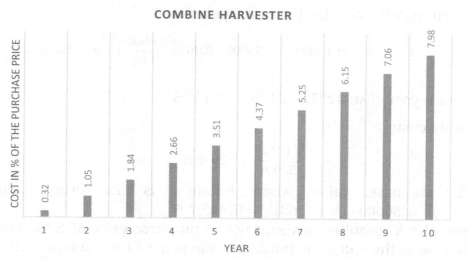

FIGURE 43 The yearly repair and maintenance cost expressed as a percentage of the purchase price for a combine (self-propelled) harvester with a useful life of 10 y and annual use of 300 h y^{-1}.

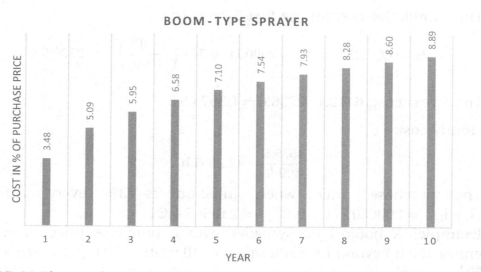

FIGURE 44 The yearly repair and maintenance cost expressed as a percentage of the purchase price for a boom-type sprayer with useful life of 10 y and annual use of 150 h y^{-1}.

The cumulative cost for the first 6 years is:

$$RM_{ac} = Q_o \cdot RF1 \cdot U_{ac}^{RF2} = 50000 \cdot 0.007 \left(\frac{6 \cdot 500}{1000} \right)^{2.0} = 3150 \text{ €}$$

Seventh year cost: $4287.5 - 3150 = $ €1137.5.

Hourly cost:

$$\frac{1137.5}{500 \, h} = 2.28 \text{ € h}^{-1}$$

The purchase value when inflation is 3% every year is $Q_o(1 + i_i)^n = 50000 \cdot (1 + 0,03)^7 = 61493.7 \text{ €}$

Example: A combine harvester with a purchase price of €200000. The combine is at the end of its third year and is used for 300 h per year. The repair and maintenance cost is detailed below.

The cumulative cost for the first 3 years is:

$$RM_{ac} = Q_o \cdot RF1 \cdot U_{ac}^{RF2} = 200000 \cdot 0.04 \cdot \left(\frac{3 \cdot 300}{1000} \right)^{2.1} = 6412.1 \text{ €}$$

The cumulative cost for the first 2 years is:

$$RM_{ac} = Q_o \cdot RF1 \cdot U_{ac}^{RF2} = 200000 \cdot 0.04 \cdot \left(\frac{2 \cdot 300}{1000} \right)^{2.1} = 2736.6 \text{ €}$$

Third year cost: $6412.1 - 2736.6 = $ €3675.5

Hourly cost:

$$\frac{3675.5}{300 \, h} = 12.25 \text{ € h}^{-1}$$

The purchase price when inflation is 3% every year is $Q_o(1 + i_i)^n = 200000 \cdot (1 + 0,03)^3 = 218545.4 \text{ €}$.

Example: A boom-type spreader with a purchase price of €5000 operates 150 h beyond its economic life (10 years—150 h per year), giving 1650 h in total.

The cumulative repair and maintenance cost is:

$$RM_{ac} = Q_o \cdot RF1 \cdot U_{ac}^{RF2} = 5000 \cdot 0.41 \cdot \left(\frac{1650}{1000} \right)^{1.3} = 3930.8 \text{ €}$$

Another way to calculate the repair and maintenance cost is by 80% of the purchase price.

The hourly cost is:

$$\frac{3930.8}{1650\,h} = 2.4\ \text{€}\ h^{-1}$$

After the end of the economic life the hourly cost will be calculated at 2.4 € h^{-1} for 150 h, and the repair and maintenance cost will be:

$$RM_{ac} = 150\,h \cdot 2.4\ \text{€}\ h^{-1} = 360\ \text{€}$$

4.1.2.2 *Fuels and Lubricants*
4.1.2.2.1 FUELS

Comparing a single machine with multiple machines, fuel consumption should be taken into consideration. The consumption of an additional liter of fuel per hour costs a lot over a long period of time. The greater the power of a machine, the more fuel it needs per hour. However, the machine completes more work in the same time. Average fuel consumption, in liters per hour, for farm tractors on a year-round basis without reference to any specific implement can be estimated with the following equation (Fig. 45).

Specific fuel consumption[71]

Diesel: $SFC_d = 2.64 \cdot P_u + 3.91 - 0.203 \cdot \sqrt{738 \cdot P_u + 173}\ \ (l\ kW^{-1}h^{-1})$

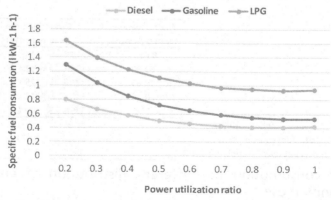

FIGURE 45 Specific fuel consumption according to the power utilization ratio with different fuel (diesel, gasoline, LPG).

Table 22 Specific Fuel Consumption (l kWh^{-1})

Load % max		Diesel		
(Power % max)	Gasoline Engines	Atmospheric	Turbo	Turbo–Intercooler
100	0.461	0.345	0.325	0.323
80	0.510	0.352	0.354	0.350
60	0.613	0.384	0.392	0.386
40	0.781	0.469	0.476	0.465
20	1.205	0.725	0.735	0.704

Gasoline: $SFC_g = 2.74 \cdot P_u + 3.15 - 0.203 \cdot \sqrt{697 \cdot P_u}\ \left(l\,kW^{-1}h^{-1}\right)$

LPG[b]: $SFC_{lpg} = 2.69 \cdot P_u + 3.41 - 0.203 \cdot \sqrt{646 \cdot P_u}\left(l\,kW^{-1}h^{-1}\right)$

where P_u is the power utilization ratio given by the ratio between the equivalent Power Take-Off (PTO) power requirements. P_{PTOeq} (kW), which represents the machine power output required either for PTO or for draught power and the maximum PTO power. P_{PTOmax} (kW), $\left(P_u = \frac{P_{PTOeq}}{P_{PTOmax}}\right)$.

The fuel consumption (FC) then is given by:

$$FC = SFC_{index} \cdot P_u \cdot P_{PTOmax} \quad for \quad index \in \{d.g.lpg\}$$

where P_{PTO} is the maximum PTO power (kW).

It is known that these equations give fuel consumption about 15% higher to cover losses due to wear and incorrect settings.

According to data from the University of Nebraska, the fuel consumption for specific field operations can be calculated for loads of 20%–100% of maximum (Table 22; Fig. 46).

Engine loading (% max power)	Operating time (% annual use)		
80-100	17%	15%	20%
60-80	24%	32%	34%
40-60	23%	25%	23%
20-40	18%	18%	18%
0-20	19%	10%	5%
Power utilization (%)	50.32	55	59
Specific fuel consumption (l/kWh)	0.502	0.479	0.460
Fuel consumtion per max PTO power (l/kW)	0.253	0.262	0.272

FIGURE 46 Fuel consumption for different distribution of engine loadings during tractor annual use.

[b]Liquefied petroleum gas.

Example: A tractor with a 100 kW diesel turbo diesel engine works with load of 80% of maximum. The fuel consumption (P_{PTO}) is 100 kW·0.6·0.354 (l kWh^{-1}) = 21.24 l h^{-1}.

4.1.2.2.2   LUBRICANTS

The most appropriate method to calculate the lubricant cost is calendar recording. If there is no data, the farmer can use the equations from the ASAE Standards.[74]

The lubrication consumption for engines of different types is given below.

Diesel: $LC_d = 0.00059 P_{max} + 0.02169 \left(l\, h^{-1} \right)$
Gasoline: $LC_g = 0.000566 P_{max} + 0.02487 \left(l\, h^{-1} \right)^1$
LPG: $LC_d = 0.00041 P_{max} + 0.02 \left(l\, h^{-1} \right)$

where P_{max} is the lubrication consumption maximum engine power (kW).

The consumption of the lubricant is based on oil changes per 100 h, it ranges from 0.0378 to 0.0946 l h^{-1} and it depends from the engine carter capacity. Generally, the cost of the lublicant can be calculated as 10% of the fuel cost.

4.1.2.3   Labor Cost

For the annual labor cost, the cost of circulating capital has to be taken into account (this also applies for the cost of fuel, lubricants, repair, and maintenance).

$$Lc = \text{annual}_{cost} \cdot i_r \cdot 0.5$$

The factor "0.5" derives from the fact that the interest on circulating capital is considered only for 6 months of the year, since operations are carried out throughout a year-long period.

Labor cost is also an important consideration in comparing ownership to custom hiring. Actual hours of labor usually exceed field machine time by 10%−20%, because of travel time and the time required to lubricate and service machines. Consequently, labor costs can be estimated by multiplying the labor wage rate by 1.1 or 1.2.

Because different sized machines require different amounts of labor to accomplish such tasks as planting or harvesting, it is important to consider labor costs in machinery analysis. Also, different wage rates can apply for operations requiring different levels of operator skill (Figs. 47−49).

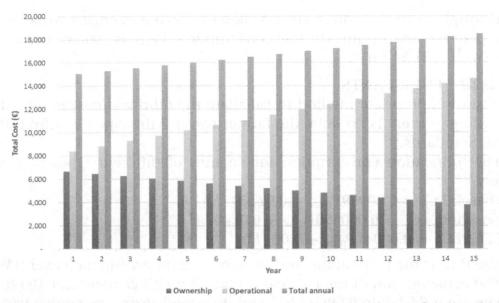

FIGURE 47 The ownership, operational, and total annual cost for the useful life of the selected example.

Example: A sample problem is used to illustrate the calculations. This example calculates the total cost with the depreciation (straight-line) method at the end of the third year.

The example uses a 60 kW diesel power tractor with the data shown below.

Purchase Value (Q_0)	€50000
Salvage value (Q_s)	10% Q_0
Useful life (n)	15 y
Annual use	800 h
Inflation rate	3% (considered fixed for the useful life period)
Interest rate	9%
Machine type	2 WD tractor ($RF1 = 0.007. RF2 = 2$)
Rated engine power P_{max}	60 kW
Maximum PTO power P_{PTOmax}	50 kW
Power utilization ratio P_u	55%
Diesel cost	0.80 € h^{-1}
Lubrication	10% of diesel cost
Labor hourly rate	5 € h^{-1}
Housing coefficient	0.75% Q_0
Insurance coefficient	1% Q_0

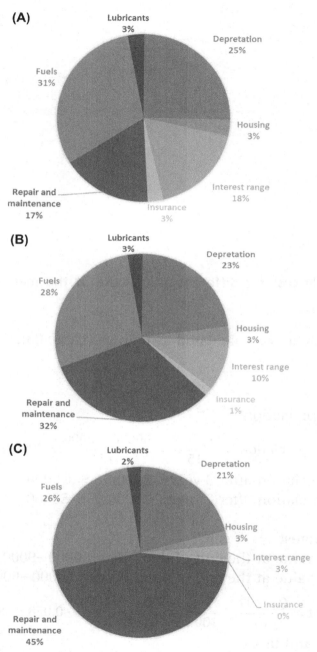

FIGURE 48 The total cost distribution (excluding labor cost) for (A) 5th year of use; (B) 10th year of use; (C) 15th year of use.

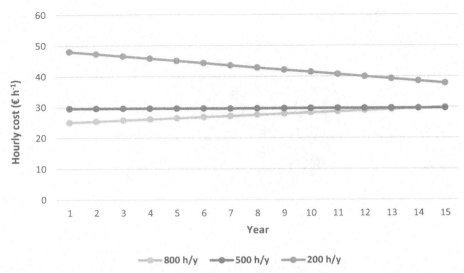

FIGURE 49 Hourly cost for different annual use of the machine throughout its useful life.

The total cost can be calculated as follows (Table 23).

1. Fixed cost.

a. Annual depreciation

$$\text{Depreciation} = \frac{Q_o - Q_s}{15} = \frac{50000 - 5000}{15} = 3000 \text{ €}$$

Total depreciation after 3 years: €3000·3 = €9000
Total depreciation after 2 years: €3000·2 = €6000

b. Capital interest
Tractor value at the end of third year: 50000−9000 = €41000
Tractor value at the end of second year: 50000−6000 = €44000

$$\text{Interest} = \frac{Q_o + Q_s}{15} \cdot \frac{i_r}{100} = \frac{44000 + 41000}{2} \cdot 0.058 = 2476 \text{ €}$$

c. Insurance and taxes

Insurance & Taxes at the end of the 3rd year = 1% of Q_3 = 0.01 · 41000 = 410 €

Table 23 Tractor Cost of Use for Different Annual Uses

Year	1	2	3	4	5	6	7	8	9	10	11	12	13	14	15
Depreciation	3000	3000	3000	3000	3000	3000	3000	3000	3000	3000	3000	3000	3000	3000	3000
Housing	375	375	375	375	375	375	375	375	375	375	375	375	375	375	375
Interest charge	2825	2650	2476	2301	2126	1951	1777	1602	1427	1252	1078	903	728	553	379
Insurance	470	440	410	380	350	320	290	260	230	200	170	140	110	80	50
Ownership	6670	6465	6261	6056	5851	5646	5442	5237	5032	4827	4623	4418	4213	4008	3804
Ownership hourly	8.34	8.08	7.83	7.57	7.31	7.06	6.80	6.55	6.29	6.03	5.78	5.52	5.27	5.01	4.75
Repair and maintenance	224	672	1120	1568	2016	2464	2912	3360	3808	4256	4704	5152	5600	6048	6496
Fuels	3679	3679	3679	3679	3679	3679	3679	3679	3679	3679	3679	3679	3679	3679	3679
Lubricants	368	368	368	368	368	368	368	368	368	368	368	368	368	368	368
Labor cost	4000	4000	4000	4000	4000	4000	4000	4000	4000	4000	4000	4000	4000	4000	4000
Interest on capital	117	117	117	117	117	117	117	117	117	117	117	117	117	117	117
Operational	8387	8835	9283	9731	10179	10627	11075	11523	11971	12419	12867	13315	13763	14211	14,659
Operational hourly	10.48	11.04	11.60	12.16	12.72	13.28	13.84	14.40	14.96	15.52	16.08	16.64	17.20	17.76	18.32
Total	15057	15301	15,544	15787	16030	16274	16517	16760	17003	17246	17,490	17733	17976	18219	18463
Total per hour	18.82	19.13	19.43	19.73	20.04	20.34	20.65	20.95	21.25	21.56	21.86	22.17	22.47	22.77	23.08

d. Housing

Housing at the end of the 3rd year $= 0.75\%$ of $Q_3 = 0.0075 \cdot 41000 = 375$ €

Total fixed cost at the end of third year $=$ €6261
Total fixed cost $= 7.83$ € h^{-1}

2. Operating cost.

a. Repair and maintenance
  i. Cumulatively for 3 years:

$$RM_{ac} = 50000 \cdot 0.007 \cdot \left(\frac{3 \cdot 800}{1000}\right)^{2.0} = 2016 \text{ €}$$

Cumulatively for 2 years:

$$RM_{ac} = 50000 \cdot 0.007 \cdot \left(\frac{2 \cdot 800}{1000}\right)^{2.0} = 896 \text{ €}$$

Costs of third year: $2016 - 896 = $ €1120
  ii. Circulating capital interest: $1120 \cdot 0.058 \cdot 0.5 = $ €33.6
  iii. Labor cost for maintenance and repairs: $0.05 \cdot 800 \text{ h} \cdot 5$ €
  $h^{-1} = $ €200
Total cost (i + ii + iii) = 1353.6 €
Total cost = 1.7 € h⁻¹

b. Fuel cost

Specific fuel conumption $= 2.64 \cdot 0.55 + 3.91 - 0.203 \cdot \sqrt{738 \cdot 0.55 + 173}$
$$= 0.478 \ l \ kW^{-1}h^{-1}$$

Fuel consuption $= 2.64 \cdot 0.55 + 3.91 - 0.203 \cdot \sqrt{738 \cdot 0.55 + 173} \cdot 0.55 \cdot 0.5$
$$= 13.138 \ l \ h^{-1}$$

Annual fuel consumption $= 13138 \ l \ h^{-1} \cdot 12 \ years \cdot \dfrac{1000}{15} \ years = 10511 \ l$

Fuel $= 10511 \ l \cdot 0.35 \text{ € } l^{-1} = 3679$ €

c. Lubrication cost

10% of fuel cost

$$Lubrication \cos t = 0.1 \cdot 3679 \, € = 368 \, €$$

d. Labor cost

i. *Labour* $\cos t = $ *Labor hourly rate* $(€ \, h^{-1}) \cdot$ *Annual use* (h)

$$Labour \cos t = 5 \, € \, h^{-1} \cdot 800 \, h = 4000 \, €$$

ii. Circulating capital interest: $4000 \cdot 0.058 \cdot 0.5 = €117€$

Total cost: $€4000 + €117 = €4117$

Hourly cost: $€4117/800 \, h = 5.15 \, € \, h^{-1}$.

Total $(a + b + c + d)$: $€9283$

Total costs (fixed and variable): $€15544$

Total hourly cost: $€15544/800 \, h = €19.43 \, h^{-1}$

4.1.3 Approximate Cost Estimation Method

The analytical method gives the most accurate estimation of cost. However, if the available data is insufficient or high precision is not required, it is possible to estimate the cost with simpler methods. The simplest method for fixed cost estimation is to calculate it as a percentage of the machine purchase price (Q_0). Depreciation is calculated as stable per year across the economic life, and the residual value is 10% of the initial value. According to this:

$$D = \frac{Q_0 - 0.1 \cdot Q_o}{n} \quad \forall i \in \{1....n\}$$

If n = 15 years:

$$D = \frac{Q_0 - 0.1 \cdot Q_o}{15} = 0.06 \cdot Q_0 \; (= 6\% \cdot Q_0)$$

If n = 10 years:

$$D = \frac{Q_0 - 0.1 \cdot Q_o}{10} = 0.09 \cdot Q_0 \; (= 9\% \cdot Q_0)$$

The interest on capital is calculated as stable throughout the economic life and it is given by the equation:

$$IC = \frac{Q_o + 0.1 \cdot Q_o}{2} \varepsilon_r$$

For $\varepsilon_r = 6\%$

$$IC = \frac{Q_o + 0.1 \cdot Q_o}{2} \cdot 0.06 = 0.033 \cdot Q_o \ (= 3.3\% \ Q_o)$$

The total annual machine use (t) can be calculated from the yearly total cultivation surface and the real field efficiency (C_E):

$$t = \frac{10 \cdot A}{C_E} = \frac{10 \cdot A}{S \cdot W \cdot E_f} \ (h)$$

The annual cost of agricultural machinery use is given by the equation:

$$AC = FC \cdot Q_O + \frac{10 \cdot A}{S \cdot W \cdot E_f} \cdot (RMC \cdot Q_O + E + F + L + T)$$

where AC = annual cost (€); FC = fixed costs as a percentage of the machine purchase price; Q_o = purchase price (€); A = operation area (ha); S = field speed (Km h^{-1}); W = working width (machine size) (m); E_f = field efficiency (decimal); RMC = repair/maintenance costs, as a percentage of Q_o h^{-1}; E = hourly labor cost (€ h^{-1}); F = hourly fuel cost (€ h^{-1}); L = hourly lubricant cost (€ h^{-1})

Example: Calculate the annual cost of a 60 kW four-wheel drive (4WD) tractor with Q_o = €50000, t = 800 h per year, fixed costs = 0.11 Q_o, repair and maintenance costs for all the economic life (16000 h) is 80% Q_o therefore RMC=0.005% Qo h^{-1} = 0.00005 Q_o h^{-1}, hourly labor cost = 4.5 € h^{-1}, and hourly fuel cost = 15.46 $l \, h^{-1}$.

Annual cost: $AC = FC \cdot Q_O + t \cdot (RMC \cdot Q_O + E + 1.1 \cdot K)$

$AC = 0.11 \cdot 50000 + 800 \cdot (0.00005 \cdot 50000 + 4.5 + 1.1 \cdot 15.46) = 24704.8$ €

Hourly annual cost: €24704.8/800 h = 30.9 € h^{-1}

Example: Calculate the annual cost of a 60 kW 4WD tractor/plow system. For the plow: Q_o = €2500, A = 20 ha, field speed S = 7 km h^{-1}, W = 1.5 m, E_f = 0.85, fixed costs = 0.1 Q_o, repair/maintenance cost = 0.0005 Q_o h^{-1}. For the tractor the cost is T = 30.9 € h^{-1} from the previous example.

In the case of a tractor/implement system the cost concerns a specific operation. In this case the total cost is the sum of the tractor cost plus the implement cost. It should be noted that the tractor hourly cost is taken from the total tractor use in the holding and not only from the tractor use with the implement.

$$AC = 2500 \cdot 0.1 + \frac{200}{7 \cdot 1.5 \cdot 0.85} \cdot (0.0005 \cdot 2500 + 30.9) = 970.4 \text{ €}$$

Cost per ha = 970.4/20 = €48.5 ha^{-1}

Cost per hour = 34.7 € h^{-1}

Example: Calculate the annual cost of a disc harrow with Q_o = €3500, A = 20 ha, field speed S = 8 km h^{-1}, W = 3 m, E_f = 0.85, fixed costs = 0.1 Q_o, repair/maintenance costs = 0.03%, Q_o h^{-1} = 0.0003 Q_o h^{-1}.

$$AC = 3000 \cdot 0.1 + \frac{200}{8 \cdot 3 \cdot 0.85} \cdot (0.0003 \cdot 3500) = 310.3 \text{ €}$$

Cost per ha= 310.3/20 = €15.5 ha^{-1}

Cost per hour = 310.3/9.8 = 31.7 € h^{-1}

Example: Calculate the annual cost of a 220 kW self-propelled combine with Q_o = €180000, A = 500 ha, field speed S = 5 km h^{-1}, W = 6 m, E_f = 0.7, fixed costs = 0.12 Q_o, repair/maintenance costs = 0.013% Q_o h^{-1} = 0.00013 Q_o h^{-1}, hourly labor cost = 4.5 € h^{-1}, hourly fuel cost = 21.5 $l\,h^{-1}$ (0.35 € l^{-1}, hourly lubricant cost = 2.15 € h^{-1}.

Hourly fuel cost + hourly lubricant cost: 1.1·Fuels = 23.65 € h^{-1}

$$AC = 180000 \cdot 0.12 + \frac{5000}{5 \cdot 6 \cdot 0.7} \cdot (0.00013 \cdot 180000 + 4.5 + 23.65) = 33873.8 \text{ €}$$

Cost per ha = 33873.8/500 = €67.7 ha^{-1}

Cost per hour = 33873.8/238 = 142.3 € h^{-1}

4.2 Indirect Cost

4.2.1 Timeliness

It has been shown that crop yield is affected by the timing of field operations. The reduction in yield due to untimely field operations is affected by many factors, such as the soil, the climate conditions, the field topography, the plant, and even the variety of the plant.

For a field operation there is usually an optimal time with respect to the value of the crop: the time when the yield is maximum. This, of course, can be determined for each crop and area only after experiments and observations. If an operation is performed earlier or later, the value of the crop may decrease due to changes in quantity and/or quality (Figs. 50 and 51). The economic consequences of performing a field

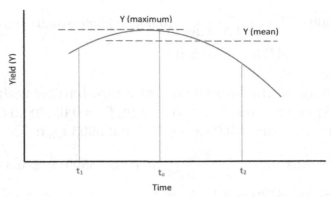

FIGURE 50 General trend of crop yield according to field operation time.

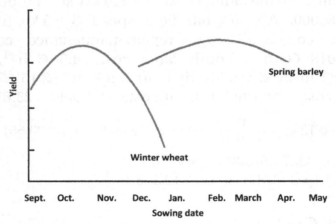

FIGURE 51 Cereal timeliness penalties as a percentage of maximum yield.[78]

operation at a nonoptimal time are called timeliness costs. If an operation starts after the optimal time, timeliness costs are incurred for the whole area before the operation starts and thereafter for a decreasing area depending on the capacity of the operation. Since these costs are partly dependent on the operation's planning and scheduling and the machine capacity, they are also called indirect machine costs. If timeliness costs are not taken into consideration there is a risk of underestimating all the costs and machinery capacity requirements.

Timeliness costs are important to consider for efficient crop management and machinery selection, particularly for crop establishment, spraying, harvesting, and soil compaction.[75,76] Significant timeliness costs can occur in regions with short sowing and harvesting periods, and since they are affected by the weather such costs are specific for regions and subject to annual variations.[77]

Except for very small holdings or large machinery sizes, in practice it is impossible to carry out all the work in the optimum period. As the operation moves away from the optimum period, the reduction in production increases, following in general terms the form of a curve (Fig. 52).

Yield losses are usually expressed as a percentage of the maximum yield. Generally, losses can be expressed as losses from early or late onset (before or after the optimum period), as well as from a balanced performance (before and after the optimum period).

When a field operation begins t_1 days earlier than the optimum period (t_0) and is completed t_2 days earlier, total production losses (percentage of the maximum) can be expressed by the equation[3]:

$$y_{total} = k_{t1} \cdot (t_2 - t_1)^{2/3}$$

where k_{t1} is the reduction coefficient for field operations before the optimal period.

Similar losses arise when the field operation begins t_1 days after the optimal period and is completed 2 days after the optimum period (t_0):

$$y_{total} = k_{t2} \cdot (t_2 - t_1)^{2/3}$$

where k_{t2} is the reduction coefficient for field operations after the optimal period.

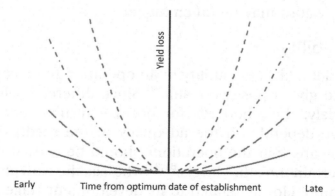

FIGURE 52 Yield losses as a function of time of crop establishment.[79]

If the field operation is programmed to begin t_1 days before the optimal period (t_o) and ends t_2 days later, the total losses are given by the equation:

$$y_{total} = \frac{1}{3} \cdot \frac{k_{t1} \cdot (t_o - t_1)^3 + k_{t2} \cdot (t_2 - t_0)^3}{t_2 - t_1}$$

With this field operation plan the total losses are the sum of the average losses when the field operation takes place earlier or later than the optimum period.

4.2.1.1 Timeliness Coefficients or Timeliness Loss Factors

When field operations are performed outside the optimum period, income is reduced. This increases the machinery cost of use when this is exclusively due to machinery weakness to operate according to the time layout or to the machinery size or to bad weather conditions or damages.

To allocate the overcharges from an out-of-time-layout field operation, the optimal operation time (t_o) for each crop and area should be found.

Example: If a 10-day delay of a field operation reduces the yield by 5%, the timeliness coefficient (C_c) is $C_c = 0.05$, for 10 days $= 0.005$ per day. If the machine operates 10 h per day the timeliness coefficient is $C_c = 0.0005$ per hour.

There are several timeliness coefficients for different field operations and different areas. These coefficients are based on a proportional relationship between the days of accelerating or delaying of the field operation and the corresponding yield reduction. The normal timeliness coefficients used are given in Table 24. In many cases, depending on the circumstances, values may be much higher.

4.2.2 Workability

Workability is defined as the ability of an operation to be carried out at a specific time to give a positive result.[80] Since different field operations' goals vary widely, the standards for positive results vary as well. For example, tillage depends on the adequacy of the seedbed to promote growth. This means that the operations should be carried out while the soil is within certain moisture levels,[81,82] whereas the workability of a crop for harvest is more closely related to its development stage.

Table 24 Timeliness Coefficients (T_c) Due to Untimely Execution for Different Field Operations[3,74]

Operation	ASAE T_c (h^{-1})	Hunt T_c (h^{-1})
Plowing	0.0000–0.0010	0.00001–0.0002
Seed drilling		0.0002–0.0006
Corn	0.0001–0.001	
Grain	0.0007–0.0008	
Cotton	0.0004–0.002	
Rice	0.001	
Row cultivation	0.001	0.001
Harvest		
Corn	0.0003	0.0003
Grain	0.0005	
Cotton	0.0002	0.0008
Rice	0.0009	
Soybean	0.0006–0.001	0.0005
Grass	0.0018	0.0010

Soil workability is directly dependent on soil moisture content, which gives us the number of available soil work days once the criterion for a work day is established. As the soil workability varies from soil to soil, machine to machine, and farm manager to farm manager, the adoption of a unique soil moisture value to differentiate between workability and nonworkability of the soil is unrealistic.

Table 25 provides information about the field operations that may be carried out under given soil moisture conditions. Generally, operations must be performed when the soil is in a semirigid state; in some cases (e.g., seeding) drier soil is required.

4.2.3 Trafficability

Trafficability is defined as the ability of soil to support a vehicle while causing negligible, or reversible, damage.[83] Damage to the soil may originate from compaction (an increase in bulk density) or deformation (changes to the structure). Both forms of damage limit the water-holding capability of the soil, limit the flow of nutrients within the soil, and cause

Table 25 Preparation in Different Soil Moisture Conditions

Soil	Field Operation					
	Compaction Correction	Plowing	Plowing Repetition	Secondary Preparation	Seeding Preparation	Seeding
Dry	Possible, difficult	Possible, difficult	Possible, difficult	Possible, difficult	Possible, difficult	Recommended
Semidry	Recommended	Recommended	Recommended	Recommended	Recommended	Recommended
Semiplastic	Dangerous	Recommended	Dangerous	Dangerous	Dangerous	Dangerous
Plastic	Destructive	Destructive	Destructive	Destructive	Destructive	Destructive

problems in root development, all of which adversely affect the final crop yield. Evaluation of soil trafficability is based on the comparison of stresses applied to the soil and on soil strength. Soil strength and deformation depend on both intrinsic properties (soil texture, mineralogy, content, and nature of organic matter) and water content.

4.2.3.1   Soil Moisture

To estimate soil moisture content on a daily basis it is necessary to know the moisture on the previous day, the precipitation, the surface runoff, the drainage, and the evapotranspiration.

The equation for soil moisture is:

$$M_s = m_p + Q_p - Q_r - Q_d - Q_e$$

where M_s is the moisture content (mm), m_p is moisture content the previous day (mm), Q_p is the precipitation (mm), Q_r is the surface runoff (mm), Q_d is the drainage (mm), and Q_e is the evapotranspiration (mm).

According to the soil and meteorological data (air temperature, wind, daily precipitation, and sunshine), altitude, and additional soil elements such as hydraulic conductivity, porosity etc., it is possible to estimate the available days for machine field operation. Meteorological data is usually available, but soil data should be identified for each area.

The global bibliography provides many tables listing possible available days and different probability levels. These tables come from real observations on the field or from simulation methods.[3,74,83] Typically, the available days for 1 or 2 weeks are given.

The farmer can choose the level of probability s/he wants. For example, if an area for soil cultivation in the second half of April has a working day probability of 0.57 at a level of 50%, this means that there are $0.57 \cdot 15 = 8.5$ appropriate days. In this case the farmer expects that in 5 out of 10 years there will be 8.5 working days. If s/he wants more safety s/he can choose 5.7 available days in 9 out of 10 years. This will be an extra cost to the farmer, because her/his machines should have greater size (productivity) in order to finish the operations in 5.7 instead of 8.5 days.

The ASAE Standards[74] gives the available days for some US states with a probability of 50% and 90%.

For soil preparation, the number of available days is also influenced by the workability criterion. So if the criterion is plastic rather than damp soil, there are more available days.[3,78]

4.2.3.2 Out-of-Time Field Operation Costs

It is possible to calculate the out-of-time operation costs of a machine when, due to smaller than required machine size or adverse weather conditions, an operation is not carried out on time, resulting in a reduction in production or a deterioration in product quality. The ASAE Standards[74] give the following equation:

$$AC_c = \frac{T_C \cdot (10 \cdot A)^2 \cdot Y \cdot V}{Z \cdot 10 \cdot C_E \cdot P_r}$$

where AC_c is the annual cost for untimely field operations, T_c is the timeliness coefficient (per hour), A is field area (ha), Y is the yield (kg ha^{-1}), V is the yield sale price (€ kg^{-1}), and C_E is the real productivity in the field (ha h^{-1}). Z is 2 when the field operation begins and ends before or after the optimum period. Z is 4 when the field operation begins at the optimum period and P is the probability of available days, decimal (the ratio of available days to the total of the calendar days)

Example: Agricultural tractor and seed drilling machine with a width of 2.5 m operate with a field speed of 8 km h^{-1} and field efficiency of 0.70. The delay of the field operation has the timeliness coefficient $T_c = 0.0003$ (per hour). A 50 ha area will be sown with corn with an expected yield of 10000 kg ha^{-1} and selling price of €0.2 kg^{-1} (with Pr = 0.5 and Z = 4).

The annual untimely field operation cost is:

$$AC_c = \frac{0.0003 \cdot 500^2 \cdot 1000 \cdot 0.2}{4 \cdot 2.5 \cdot 8 \cdot 0.7 \cdot 0.5} = 535.7 \,€\, \left(= 1.07 \,€\, acre^{-1} \right)$$

4.2.4 Reliability

As a consequence of their high productivity, the cost of damage is an important issue for many agricultural machines, particularly for specialized machines like the combine harvester. Few things cause more frustration than a functional failure in a high-performance seasonal machine, especially when the farmer faces difficult weather conditions and is behind schedule. Repair costs can be significant, and subsequent crop losses may mount up due, for example, to the delayed harvest.[84] Studies have shown that machine downtime reduces the potential in-season harvesting area by 20% and increases costs by 6%–7%.[85] In terms of statistical reliability measures, ASAE[72] reports that the probability of at least one failure in a machine system per year varies between 44% and 78% depending on the size of the crop area.

The prediction of agricultural machinery performance is an important aspect of machinery management.[3] A key parameter is the probability of equipment breakdown, and the machine manager must evaluate this in her/his efforts to optimize machinery operations and identify the most reliable machines. Machines can be highly reliable but costly or less reliable and inexpensive, indicating a trade-off and that some probability of equipment breakdown must be taken into account.[86] The operational performance of farm machinery must be quantified to be able to select and plan the use of equipment under different conditions. However, most of the models developed for machinery management have used the deterministic notion of performance measures like working width, working speed, loading capacity, etc.[87] Probabilistic measures like reliability and availability have not received as much attention, despite the fact that a number of studies have shown that machinery systems operate in a stochastic environment. So models of the management process have to include probabilistic measures to capture all implications in the system. However, the difficulty in dealing with probabilistic measures like availability and reliability concerns their practical implementation in terms of quantification and operational usage.

Machine downtime, i.e., the time lost due to breakdowns, has to be taken into account in the planning stage, otherwise failures can significantly affect machine utility in intensive operation periods such as the harvesting season. However, agricultural or industrial reliability studies as an integral part of planning efforts have not received much attention in the past. One of the main reasons has been the difficulty in collecting useful data, due to the fact that agricultural machines change continuously with respect to type and attributes, thus it is difficult to acquire reliability measures on new models and types, and also agricultural machines operate very differently because the crops processed by the machines vary in a number of physical attributes. Consequently, agricultural reliability data varies significantly and has to be regarded as random variables.

Observation data shows that in many agricultural machines damage is independent of the operating time, while in others there is increased damage as time passes (ASAE Standards).[74]

Table 26 presents the breakdown possibility for tractors and implements combined (ASAE Standards).[88]

Breakdown probabilities for machine systems increase as the size of the farm increases. Table 27 presents the probability of at least one failure per year and the reliability of tractor/machine systems per year according to crop area.

Table 26 Breakdown Possibility for Tractors and Implements Combined

Operation	Breakdown Time (h year^{-1})	Breakdown Probability Per 40 ha	Reliability Per 40 ha
Tillage	13.6	0.109	0.89
Planting corn	5.3	0.133	0.87
Planting soybeans	3.7	0.102	0.90
Row cultivation	5.6	0.045	0.96
Harvest soybeans	8.2	0.363	0.64
Harvest corn	12.3	0.323	0.68

Table 27 Breakdown Probability of at Least One Failure and Reliability of Tractor and Implement Combined Per Year[72]

Crop Area (ha)	Probability of at Least One Failure Per Year	Reliability of Tractor and Implement Combined Per Year
0–80	0.435	0.56
80–160	0.632	0.30
160–240	0.713	0.29
>240	0.780	0.22

Downtime and reliability appear to be independent of use for some machines, while others show an increase with accumulated use. In moldboard plows an average 1 h downtime for each 400 ha of use is observed, in row planters there is an average 1 h downtime for each 250 ha of use, and self-propelled combines have very low downtime for the first 365 ha of use. The accumulated hours of downtime depend upon the accumulated hours of use.

In spark ignition engines the total downtime hours (AD) are calculated by the equation:

$$AD = 0.0000021 \cdot t^{1.9946}$$

In diesel engines the accumulated hours of downtime (AD) are calculated by the equation:

$$AD = 0.000323 \cdot t^{1.4173}$$

where *t* is the total operating time in hours.

Example: Consider a tractor with 500 operational hours a year. The downtime hours in the seventh year are as follows.

Accumulated   hours   of   downtime   after   7 years: $AD = 0.000323 \cdot 3500^{1.4173} = 34$ h.

Accumulated   hours   of   downtime   after   6 years: $AD = 0.000323 \cdot 3000^{1.4173} = 27$ h.$Q_s = 0.1 \cdot Q_0$

The downtime hours at the seventh year are $34 - 27 = 7$ h for 500 operational hours per year. In tractors, research has shown that almost 50% of damage is found in the transmission system.[66] From the above analysis, it seems that the cost of possible damage is significantly less than the cost due to other causes. Sometimes, however, due to poor design and possible lack of spare parts, machines, especially tractors, may be out of service for a long time. If this happens in critical periods (sowing, etc.) it is possible, in extreme cases, to lose the whole growing season.

Choosing a Machinery System

5.1 Tractor Selection

Tractor selection is one of the most difficult decisions for a farmer, and will affect almost all the equipment of the farm for a long time and the cost of the products. The most significant factors in tractor selection are:

- farm holding and crop type
- soil composition and morphology
- size and number of farmland parcels
- farm weather conditions
- timeliness of operations
- projected changes of the crops
- ability to find labor and its cost
- purchase price of the tractor and implements
- current popularity of different tractors in the area
- existence of proper organized maintenance services from the tractor's agents.
- operator safety and comfort.

5.1.1 Tractor Required Power Calculation Based on Minimum Cost

For the required power calculation, the equation of the annual cost must first be compiled and then minimized with power differentiation. The annual cost of use, as determined in Chapter 4 based on the approximate method, is given by the equation:

$$AC = FC \cdot Q_O + \frac{10 \cdot A}{S \cdot W \cdot E_f} \cdot (RMC \cdot Q_O + E + F + L)$$

where AC, annual cost (€); FC, fixed costs as a percentage of the machine purchase price ; Q_o, purchase price (€); A, operation area (ha); S, field speed (Km h^{-1}); W, working width (machine size) (m); E_f, field efficiency (decimal); RMC, repair/maintenance costs as a percentage of Q_o h^{-1}; E, labor cost (€ h^{-1}); F, fuel cost (€ h^{-1}); L, lubricant cost (€ h^{-1}).

The factor A/S·W·E$_f$ represents the total machine working time (t) in hours (h).

If the purchase price Q_o is expressed based on the power P (P$_{PTO}$), it will be $Q_o = q \cdot P$ (€), where q is the value per PTO (Power Take Off) power unit (€ kW^{-1}).

The cost of fuel can be expressed as a function of the power (kW) of the specific consumption (kg kW^{-1}h^{-1} or l kW^{-1}h^{-1}) and fuel price (€ kg^{-1} ή € l^{-1}) as $K = \kappa \cdot P$ (€ h^{-1}), where $\kappa = $ € kW^{-1}.

Similar to fuel, the lubricant cost is $L = \lambda \cdot P$, where $\lambda = $ € kW^{-1}. The relation between the tractor annual cost and its power will therefore be:

$$AC = FC \cdot q \cdot P + t \cdot (RMC \cdot q \cdot P + E + \kappa \cdot P + \lambda \cdot P)$$

Usually the lubricant cost is around 10% of fuel cost. With this assumption, the annual cost (AC) is:

$$AC = FC \cdot q \cdot P + t \cdot (RMC \cdot q \cdot P + E + 1.1 \cdot \kappa \cdot P)$$

Because the energy required to perform the tasks is $E_v = P \cdot t$, replacing time t with the equivalent of $t = E_N/P$:

$$AC = FC \cdot q \cdot P + RMC \cdot q \cdot E_N + \frac{E_N \cdot E}{P} + 1.1 \cdot \kappa \cdot E_N$$

This final cost equation is a function of energy and power. To find the power that gives the minimum cost, the partial differential of power relation (P) is taken. The equation has a minimum (minimum P$_{PTO}$ value), which is found if the partial differential in power equals zero (0):

$$\frac{\partial AC}{\partial P} = FC \cdot q - \frac{E_N \cdot E}{P^2} = 0$$

so:

$$P = \sqrt{\frac{E_N \cdot E}{FC \cdot q}}$$

From this equation, it appears that the size of the tractor is affected by the energy required to carry out the work, the size of the workforce, the cost per unit of power, and the fixed costs. It is worth noting that maintenance/repair and fuel/lubricant costs are not involved, because they are functions of power.

More specifically, energy is affected by the size of the holding, the crops and the area they cover, the number of crop interventions, the tools and machinery used, the mechanical composition of the soil, the slope, the speed and depth of work, and real field performance.

Fixed costs are mainly dependent on depreciation, interest on capital, and housing and insurance costs. These costs are mainly affected by the purchase price, the duration of the economic life, and the interest rates on the funds. It is thus observed that almost all the theoretical factors developed previously are involved in the costing equation and the final choice of size.

The final equation shows that if the needs of the farm are greater (more energy is required), the size of the tractor should be greater as well, which is considered to be self-evident. If there is a high labor requirement the tractor should be bigger, to complete operations sooner and reduce labor costs. In contrast, the greater the buying cost of the tractor, the higher its interest rate or the shortest its economic life, the smaller the size of the tractor should be; so that when its annual use is increased, the fixed costs per hour are reduced and subsequently so it is the hourly total cost.

For better understanding of the calculation of the required power, the following example of an agricultural holding is given. For reasons of convenience, the example refers to a simple farm of 22 ha under cultivation (Table 28).

The other necessary elements are as follows.

The fixed costs of the tractor amount to 11% of the purchase price.
The price of tractors per kWh in PTO is €700.
The operator cost is 4.5 € h^{-1}.

Table 28 Type of Crops and Their Area

Crop	Crop Area (ha)	Soil
Grain	5	Medium mechanical composition
Cotton	10	Medium mechanical composition
Corn	7	Medium mechanical composition

The average distance of the fields from the house is estimated at 2 km. To calculate the required tractor power, it is necessary to draw up work logs and calculate the required energy.

5.1.1.1 Draw the Task Log
The selection procedure for the tractor (appropriate tractor size) starts from detailed logging of the work to be carried out on the farm in which the tractor will be involved. Essentially, this is about creating a so-called work diary. Based on the crops, conditions, and habits of the producer, a calendar is created (Table 29).

5.1.1.2 Calculation of Required Energy
The next step in the calculations is to estimate the energy required to perform the crop operations in Table 29. The calculations are based on the data in Appendix Table I. The total energy requirements of the holding are shown in Table 30.

The power calculations in Table 30 are based on the data in Table I in the Appendix and facilitate further calculations made per meter of

Table 29 Crop Workbook

| | Number of Interventions | | | | | |
| | Crops | | | ha | | |
Operations	Grain	Cotton	Corn	Grain (5)	Cotton (10)	Corn (7)
Plowing	1	1	1	5	10	7
Disk harrowing	1	1	2	5	10	14
Cultivating		3	2		30	14
Rotary tilling		1			10	
Basic fertilizing	1	1	1	5	10	7
Harrowing	1	1	1	5	10	7
Seeding and fertilizing	1	1	1	5	10	7
Row cultivation		2	2		20	14
Spraying	1	4	2	5	40	14
Straw cutting		1	1		10	7
Straw baling	1			5		
Irrigation		5	5		50	35
Movements	9	30	20			
Transport	1	10	1			

Table 30 Traction and Power Requirements for Different Tasks

Operations	Traction Power (kW) Soil Medium	Heavy	Operating Speed (km h^{-1})	Field Efficiency (%)	Energy (kWh ha^{-1}) Soil Medium	Heavy
Plowing	36.8	52.6	7.0	80	65.8	93.9
Disk harrowing	14.4	16.4	10.0	80	18.0	20.5
Cultivating	4.4	5.2	8.0	85	6.5	7.7
Rotary tilling	18.3	23.7	5.0	85	43.1	55.8
Basic fertilizing	2.1	2.1	11.0	70	2.7	2.7
Harrowing	1.8	1.8	11.0	85	2.0	2.0
Seeding and fertilizing	5.0	5.0	9.0	65	8.5	8.5
Grain seeding and fertilizing	4.4	4.4	8.0	70	7.9	7.9
Row cultivation	2.2	2.6	8.0	80	3.4	4.1
Spraying	2.3	2.3	10.0	65	3.5	3.5
Straw cutting	11.0	11.0	8.0	80	17.2	17.2
Straw baling	31.5	31.5	8.0	65	60.6	60.6
Movements	11.6	11.6	20.0		0.58 (kWh km^{-1})	0.58 (kWh km^{-1})
Transport	38.9	38.9	20.0		1.95 (kWh km^{-1})	1.95 (kWh km^{-1})
Transport					0.195 (kWh km^{-1}.t^{-1})	0.195 (kWh km^{-1}.t^{-1})

amplitude. Since the mechanical composition of the soil has a significant impact on the required power, the power and energy required for comparisons have also been calculated for heavy soils.

For a better understanding of the calculations, the following example is given. For soil treatment with a moldboard plow, Appendix Table I gives parameter values of the machine and the soil, from which the total resistance can be calculated[74]:

$$D = F_i \cdot \left[A + B \cdot (S) + C \cdot (S)^2 \right] \cdot W \cdot T$$

where D, implement draft (N); A, B, C, soil factors (Table I in Appendix); F_i, soil parameter (Table I in Appendix); S, field speed $(7\,km^{-1})$; W, machine width (m) (for example 1 m); T, tillage depth (for example 30 m)

After replacing the values:

$$D = 0.7 \cdot (652 + 0 \cdot 7 + 5.1 \cdot 7^2) \cdot 1 \cdot 30 = 18940\,N$$

Power is calculated as follows:

$$P = \frac{D \cdot S}{1000}\,(kW)$$

with value replacements:

$$P = \frac{18940 \cdot \dfrac{7000}{3600}}{1000} = 36.8\,kW$$

The energy is calculated by the equation:

$$E_N = P \cdot t\,(kWh)$$

The time required to plow one ha is calculated from the actual field yield:

$$C_E = W \cdot S \cdot E_f$$

By replacing values:

$$C_E = 1 \cdot 7 \cdot 0.8 = 0.56\,ha\,h^{-1} \quad or\ 1/C_E = 1/0.56 = 1.786\,h\,ha^{-1}$$

The energy per ha is:

$$E_N = 36.8\,(kW) \cdot 1.786(h\,ha^{-1}) = 65.8 kWh\,ha^{-1}$$

For PTO powered implements, Appendix Table II gives parameters for the below equation[74]:

$$P_{PTO} = a + b \cdot w + c \cdot F\,(kW)$$

where a, b, c, parameters from Table II in the Appendix. n, working width (m); P, material entering the machine $(t\,h^{-1})$.

In many cases, beyond PTO power, power is also required to pull implements (Table II in the Appendix gives additional data). These are usually wheel-mounted and rolling resistance tools.

Special attention is needed in the calculation of energy for movement and transportation. The calculation requires the total weight of the tractor and implements or the tractor and platform, the rolling resistance coefficient (0.05–0.12, with an average for rural roads of 0.07), and the speed of movement (20 km h^{-1}).

Based on the above, a total weight of 3000 kg with a rolling resistance coefficient of 0.07 becomes 210 kg = 2100 N = 2.1 kN.

$$P = \frac{2100 \cdot \frac{20000}{3600}}{1000} = 11.6 \ (kW)$$

If the speed is 20 rpm, the time for 1 km is t = 1/20 = 0.05 h. The energy for moving 1 km will therefore be:

$$E_N = 11.6 \cdot 0.05 = 0.58 \ kWh \ km^{-1}$$

For transportation the total load was calculated as 10000 kg (tractor and platform). Based on a speed of 20 km h^{-1} and a rolling resistance factor of 0.07, the power is:

$$P = 7(kN) \cdot 20(km \ h^{-1})/3.6 = 38.9 \ kW$$

and the energy:

$$E_N = 38.9 \ (kW) \cdot 0.05(km \ h^{-1}) = 1.95 kWh \ km^{-1}$$

If divided by the total load (10000 kg), the energy becomes:

$$E_N = 0.195 \ kWh \ km^{-1} \cdot t^{-1}$$

Based on Tables 29 and 30, Table 31 shows the requirements of the three crops in energy for each work separately, but also for the total. Movement and transportation energy is expressed per ha. For movements, the energy is calculated as follows.

Grain: 9·4 km (transition and return)·0.58 kWh km^{-1} = 20.88 kWh. Per ha, 20.88/5 = 4.2 kWh ha^{-1}.

Cotton: 30·4·0.58 = 69.6 kWh. Per ha, 69.6/10 = 6.96 kWh ha^{-1}.

In addition, the returns of the empty platform that carried cotton: 10 movements·2 km·0.58 = 11.6 kWh. Per ha, 1.2 kWh ha^{-1}. Total 7 + 1.2 = 8.2 kWh ha^{-1}.

Corn: 20·4·0.58 = 46.4 kWh. Per ha, 46.4/7 = 6.6 kWh ha^{-1}.

For transport, the calculation is as follows.

Grain: 4 t·4 km (transit and return)·0.195 kWh km^{-1}.t^{-1} = 3.12 kWh. Per ha, 3.12/5 = 0.63 kWh ha^{-1}.

Table 31 Energy Requirements of Different Operations in Three Crops

Operations	Grain	Cotton	Corn	Grain	Cotton	Corn
	Energy (kWh ha^{-1})			Energy (kWh ha^{-1})		
	Medium Soil			Heavy Soil		
Plowing	65.8	65.8	65.8	93.9	93.9	93.9
Disk harrowing	18.0	18.0	36.1	20.5	20.5	41.0
Cultivating	0.0	19.6	13.1	0.0	23.1	15.4
Rotary tilling	0.0	43.1	0.0	0.0	55.8	0.0
Basic fertilizing	2.7	2.7	2.7	2.7	2.7	2.7
Harrowing	2.0	2.0	2.0	2.0	2.0	2.0
Seeding and fertilizing	0.0	8.5	8.5	0.0	8.5	8.5
Grain seeding and fertilizing	7.9	0.0	0.0	7.9	0.0	0.0
Row cultivation	0.0	6.9	6.9	0.0	8.1	8.1
Spraying	3.5	14.2	7.1	3.5	14.2	7.1
Straw cutting	0.0	17.2	17.2	0.0	17.2	17.2
Straw baling	60.6	0.0	0.0	60.6	0.0	0.0
Movements	4.2	8.2	6.6	4.2	8.2	6.6
Transport	3.4	17.6	0.5	3.4	17.6	0.5
Total (kWh ha^{-1})	168.1	223.7	166.3	198.8	271.8	203.0
Total (kWh)	840.60	2236.72	1164.38	993.82	2718.01	1421.33
Farm grand total (kWh)	4241.70			5133.16		

Total energy requirements amount to 4241.70 kWh for medium soils or 5133.16 kWh for heavy soils.

Additional straw delivery: 3 (transports)·6 (t average transition and return weight)·4 km·0.195 kWh km^{-1}.t^{-1} = 14 kWh. Per ha, 14/5 = 2.8 kWh ha^{-1}. Total: 0.63+2.8=3.43 kWh ha^{-1}.

Cotton: 10 (transports)·2 km·45 (t production and supplies)· 0.195 kWh km^{-1}.t^{-1} = 175 kWh. Per ha, 175/10 = 17.5 kWh ha^{-1}.

Corn: As for grain, 3.12 kWh. Per ha, 0.45 kWh ha^{-1}.

5.1.1.3 Calculation of Required Power
According the following equation:

$$P = \sqrt{\frac{E_N \cdot E}{FC \cdot q}}$$

and with value replacements, the power is:

$$P = \sqrt{\frac{4241.70 \cdot 4.5}{0.11 \cdot 700}}$$

For heavy soils power is estimated at 17.3 kW.

The power is that required to pull the implements. This calculation is made on the basis of the requirements of the components in traction power.

For the required PTO power calculation, the corresponding PTO power must be calculated for every operation from the tractive efficiency (TE) when it is known or from the tractive and Kepner transmission (Table 32) ($P_{A\xi}/P_{PTO} = 0.94{-}0.96$).

5.1.2 Tractor's Required Power Calculation Based on Timely Completion of Operations (Optimal Cost Method)

Out-of-time field operations lead to yield reduction or product quality degradation. The decrease in income should be added to the cost of using a tractor or other self-propelled machinery.

In many countries different charge coefficients have been established for each field operation and region. The most common are presented in Table 24.

The charge annual cost is calculated from the charge coefficient[71]:

$$AC_{oc} = \frac{C_C \cdot (10 \cdot A)^2 \cdot Y \cdot V}{Z \cdot C_E}$$

where AC_{oc} is the charge annual cost due to untimely field operations, C_c is the charge coefficient (per hour), A is field area (ha), Y is yield, V is yield sale price (€ kg^{-1}), C_E is the real productivity in the field, Z is 2 when the

Table 32 Traction and Transmission Coefficients

Surface Condition	Light Traction Load	Medium Traction Load	High Traction Load Without Slippage
Concrete	0.75	0.85	0.9
Compact soil	0.6	0.75	0.8
Cultivated soil	0.4	0.55	0.65
Sand soil	0.25	0.4	0.45

field operation begins and ends before or after the optimum period, and Z is 4 when the field operation begins at the optimum period.

$$M_E = (C_C \cdot A \cdot Y \cdot V)/Z \quad (\text{€h}^{-1})$$

According to the previous example, the required power using the optimal cost method is $C_c = 0.0004$ per hour.

Grain: Yield production 4000 kg ha^{-1}. Price 0.13 € kg^{-1}

Cotton: Yield production 4000 kg ha^{-1}. Price 0.80 € kg^{-1}

Corn: Yield production 8000 kg ha^{-1}. Price 0.13 € kg^{-1}

The hourly charges (ME) with $Z = 4$ are calculated as follows.

Grain: 0.26 € h^{-1}, Cotton: 3.2 € h^{-1}, Corn: 0.73 € h^{-1}.

The total hourly charges for all the crops based on the hourly charges for each crop are ME = 4.19 (€ h^{-1}).

With $Z = 2$ the hourly charges are calculated as follows.

Grain: 0.52 € h^{-1}, Cotton: 6.4 € h^{-1}, Corn: 1.46 € h^{-1}.

The total hourly charges become ME = 8.38 (€ h^{-1})

The tractor's annual operation cost takes into account the charge for untimely operations.

$$AC = FC \cdot Q_0 + \frac{10 \cdot A}{S \cdot W \cdot E_f}(RMC \cdot Q_0 + E + F + L + M_E)$$

$$AC = FC \cdot q \cdot P + t \cdot (RMC \cdot q \cdot P + E + 1.1 \cdot \kappa \cdot P + M_E)$$

replacing time t with the equivalent of $t = E_N/P$:

$$AC = FC \cdot q \cdot P + RMC \cdot q \cdot E_N + \frac{E_N \cdot E}{P} + 1.1 \cdot \kappa \cdot E_N + \frac{M_E \cdot E_N}{P}$$

This equation depends on power. To calculate the power, we use the following equation:

$$\frac{\partial AC}{\partial P} = FC \cdot q - \frac{E_N E}{P^2} - \frac{M_E E_N}{P^2}$$

$$P = \sqrt{\frac{E_N(E + M_E)}{FC \cdot q}} \quad (kW)$$

Instead of the total hourly charges (M_E), the hourly charges of individual crops can be used:

$$AC = FC \cdot q \cdot P + t \cdot \left\{ [RMC \cdot q \cdot P + E + 1.1 \cdot \kappa \cdot P] \right.$$

$$+ \left[\frac{(C_C \cdot 10 \cdot A_{grain} \cdot Y_{grain} \cdot V_{grain})}{Z} + \frac{(C_C \cdot 10 \cdot A_{cotton} \cdot Y_{cotton} \cdot V_{cotton})}{Z} \right.$$

$$\left. \left. + \frac{(C_C \cdot 10 \cdot A_{corn} \cdot Y_{corn} \cdot V_{corn})}{Z} \right] \right\}$$

For $Z = 4$:

$$P = \sqrt{\frac{7026.25(4.5 + 4.19)}{0.11 \cdot 700}} = 28.2 \ (kW)$$

For $Z = 2$:

$$P = \sqrt{\frac{7026.25(4.5 + 8.38)}{0.11 \cdot 700}} = 34.3 \ (kW)$$

The corresponding value found using the minimum cost method is 20.3 kW$_{PTO}$.

According to this method, the total charges occured from the not on time field operations and the operations of limited time are calculated by the following equation:

$$AC_{oc} = \frac{C_C \cdot (10 \cdot A)^2 \cdot Y \cdot V}{Z \cdot p_r \cdot C_E}$$

where AC_{oc} is the charge annual cost due to untimely field operations, is the charge coefficient (per hour), A is field area (ha), Y is the yield, V is the yield sale price ($\in$ kg^{-1}), C_E is the real productivity in the field, p_r is the ratio of available days in terms of total calendar days, Z is 2 when the field operation begins and ends before or after the optimum period, and Z is 4 when the field operation begins at the optimum period.

Example: With data from the previous examples and for $Z = 4$ and $Z = 2$, the hourly charges per cultivation are calculated as follows:
Grain: 0.51 $\in$ h^{-1}. Cotton: 6.28 $\in$ h^{-1}. Corn: 1.43 $\in$ h^{-1}.
The total hourly charges for all the crops are 8.22 ($\in$ h^{-1})
With $Z = 2$ the hourly charges are calculated as:
Grain: 1.02 $\in$ h^{-1}. Cotton: 12.56 $\in$ h^{-1}. Corn: 2.86 $\in$ h^{-1}.
The total hourly charges become 16.44 ($\in$ h^{-1})

The tractor power is calculated on the best value method based on average hourly charges:

$$P = \sqrt{\frac{E_N(E + M_E)}{FC \cdot q}} \ (kW)$$

where M_E is the total hourly charges.

For $Z = 4$:

$$P = \sqrt{\frac{7026(4.5 + 8.22)}{0.11 \cdot 700}} = 34.1(kW)$$

For $Z = 2$:

$$P = \sqrt{\frac{7026(4.5 + 16.44)}{0.11 \cdot 700}} = 43.7(kW)$$

5.2 Equipment Selection

The successful selection of a tractor is complemented by choosing suitable ancillary equipment to keep the cost of use of agricultural machinery as low as possible. For equipment selection there are many methods, similar to those used for tractors.[3,66]

The simplest methods are based on the available power of the tractor or the time required to carry out the work; despite their simplicity, they have credibility. The most complex and advanced methods are based on the operation cost (tractor cost plus equipment cost), which should be minimal; these are cost-minimization methods. For the simple methods, cost is considered to be only the direct cost; in the more advanced methods both direct and indirect costs are taken into account, including inadequate execution of work and limits on time as analyzed in the tractors section 5.1. The second category of methods aims at cost optimization.[78]

5.2.1   Calculation of Implement Size Based on Available Tractor Power

By this method the choice of equipment size is based on the tractor power. The size of the equipment should be such that the resistance while operating causes the tractor to load about 75%–90% of its power, so there is capacity to handle overloads. Calculation of the required size should take into account the size of the resistance per width unit, the recommended speed in the field, and the degree of power output in the traction (or the transmission and traction coefficients). Based on the given tractor pulling power, it is easy to determine the working width (size) of the implement.

Example: An agricultural holding which has a tractor with maximum power of 80 kW$_{PTO}$ needs a plow for plowing on average mechanical grounds at a speed of 7 km h^{-1} and a depth of 35 cm. The moving resistance for the plow is 18,940 N per meter of width (for 35 mm depth); for a heavy ground the resistance reaches 27,057 N m^{-1}. The power efficiency in traction (TE) is 0.62, and the tractor load factor is 95%. The charging is chosen of high value because the operation is very heavy (35 mm depth) and of short duration.

The size of the plow is calculated as follows:

$$TE = \frac{P_{tr}}{P_{A\Xi}}$$

and

$$\frac{P_{A\Xi}}{P_{PTO}} = 0.96$$

From the above equations:

$$P_{tr} = TE \cdot P_{A\Xi} = TE \cdot P_{PTO} \cdot 0.96$$

Replacing the values:

$$P_{tr} = 80 \cdot 0.96 \cdot 0.62 = 47.62 \text{ kW}$$

With a charge of 95%:

$$P_{tr} = 47.62 \cdot 0.95 = 45.24 \text{ kW}$$

The available tractor power for plow traction is 28.27 kW.

The traction power is given by the equation:

$$P_{tr} = \frac{F \cdot u}{1000} \text{ (kW)}$$

and soil resistance $F = 18{,}940$ N m^{-1}.

With price replacement:

$$P_{tr} = \frac{F \cdot u}{1000} = \frac{18940 \cdot \dfrac{7000}{3600}}{1000} = 36.83 \text{ kW}$$

Thus for each measure of working width, power of 36.83 kW is required. Since the traction power available to the tractor on the holding is 36.83 kW, the maximum width is:

$$W = \frac{45.24}{36.83} = 0.77 \text{ m} = 122 \text{ cm}$$

The plow should have a width of 120 cm (three-furrow plow, 40 cm each).

If the data of the example in the previous section 5.1 (5 ha of wheat, 10 ha of cotton, and 7 ha of corn) is applied and one plow per year is used in the total area, on the basis of the data the total time is:

$$t = \frac{10 \cdot A}{S \cdot W \cdot E_f} = \frac{10 \cdot 22}{7 \cdot 1.2 \cdot 0.8} = 32.74 \text{ h}$$

With the plow selected, using the 80 kW tractor, whose hourly cost for a total operation on the holding of 500 h is $T = 23.16$ € h^{-1}, and with additional elements for the plow of $Q_o = 1600$ € m^{-1}, $RM = 0.0005\ Q_o$ h^{-1}, $FC = 13\%\ Q_o$, the tillage costs are:

$$AC = FC \cdot Qo + \frac{A}{S \cdot W \cdot E_f}(RM \cdot Qo + T) \quad (\text{€})$$

Replacing the values:

$$AC = 1600 \cdot 0.13 + \frac{220}{7 \cdot 1.2 \cdot 0.8}(0.0005 \cdot 1600 + 23.16) = 992.45 \text{ €}$$

Thus we have:

Cost per hour = €30.32
Cost per ha = €45.1

From the above cost of 30.32 € h^{-1}, 80% (24.26 € h^{-1}) is due to the operation of the tractor and only 20% due to the plow.

This method of calculation gives the size of implements that actually use the tractor's available power, and this is its main advantage. However, it is based on the power of the tractor, regardless of whether its design is correct or not. The method gives reliable results for implements that require great power for their operation, such as plows, disk harrows, earth tillers, heavy cultivators, and others.

5.2.2 Calculation of Implement Size Based on Available Time for Field Operations

This method calculates the implement size based on the available time for performing the tasks and the actual performance of the machinery in the field. However, special care should be taken not to carry out two or more operations simultaneously that need power from the same tractor.

The actual performance of the machinery in the field is:

$$C_E = 0.1 \cdot S \cdot W \cdot E_f \, (\text{ha h}^{-1})$$

where S = field speed (Km h^{-1}); W = machine width (m); E_f = field efficiency (decimal).

The total field area (A) that the machine works on is equal to the actual performance in the field during the operation time of the machine or:

$$A = C_E \cdot t = S \cdot W \cdot E_f \cdot t$$

and therefore the size (width) W is equal to:

$$W = \frac{10 \cdot A}{S \cdot E_f \cdot t} \, (m)$$

Example: Area of 20 ha in which cotton (10 ha) and corn (10 ha) will be cultivated twice with a disk harrow with a field speed of 8 km h^{-1}, in a time of 50 h. The field efficiency is 0.80. The size of the disk harrow will be:

$$W = \frac{2 \cdot 10 \cdot 20}{8 \cdot 0.8 \cdot 50} = 1.25 \, (m)$$

This size is the minimum to complete the work within 50 h.

Since the size has been set, the power of the tractor must be calculated. If the power is sufficient, we next determine the cost. If the power is insufficient, another solution should be found, such as an increase in time, e.g., 60 h instead of 50 h, partial use of professional machinery, or some other way. However, in all these cases the study should be completed with the cost calculation.

In the example studied the resistance of average ground on the disk harrow is 5200 N m^{-1}, thus the required power is (RP = 0.45):

$$P_{tr.} = \frac{5200 \cdot 1.25 \cdot \dfrac{8000}{3600}}{1000} = 14.4 \ (kW)$$

and

$$P_{PTO} = \frac{P_{tr.}}{0.96 \cdot TE} = \frac{14.4}{0.95 \cdot 0.45} = 33.68 (kW)$$

with a load of 80% of the maximum power we have:

$$P_{PTO} = \frac{33.68}{0.8} = 42.1 (kW)$$

The power of the tractor is 80 kW$_{PTO}$, so it is sufficient.

The cost of disk harvesting with additional elements $Q_o = 4000$ € m^{-1}, FC = 13% Q_o, RM = 0.0003 Q_o h^{-1}, and T = 23,16 € m^{-1} is:

$$AC = FC \cdot Q_o + t \cdot (RM \cdot Q_o + T)$$

Replacing with numbers, we have:

$$EK = 4000 \cdot 1.25 \cdot 0.13 + 50 \cdot (4000 \cdot 1.25 \cdot 0.0003 + 23.16) = 1883 \ (€)$$

From the above we have the following outcomes:

Cost per hour = €37.66
Cost per ha = €47

The implement size calculated using this method allows completion of the work within the time available as long as there is sufficient power for their execution. Its disadvantages are that it does not take into account the operating costs, or the power required for operation of the implements. Last but not least, there is a difficulty in estimating the time available, and this should also be taken into account.

5.2.3 Calculation of Implements Size Based on Minimum Cost Method

The calculation of implement size with reference to minimum cost is based on the corresponding method for the tractor's power calculation. In other words, the cost of the tractor and the implement is reduced at its minimum levels.

The equation for machine operation cost is:

$$AC = FC \cdot Q_o + \frac{10 \cdot A}{SW \cdot E_f} \cdot (RM \cdot Q_o + E + K + \Lambda + T_s)$$

where T_S = tractor stable cost per hour.

The purchase price Q_o can be expressed, similar to the tractor power, with reference to the value per width unit of the machine and the working width. As we already know, the size of the machine has an effect on its performance on the field, but this effect may be disregarded. Repair and maintenance costs are also proportional to the size, and can be $RM \cdot q \cdot W$. The cost of fuels and lubricants depends on the size of the machine. To simplify the calculations, this cost is considered as proportional to the size and can be calculated as $k \cdot W$ and $\lambda \cdot W$. Labor costs are independent of the size of the machine. The tractor fixed costs per hour (T_S) are considered as a function of operation time, independent of the size of the machine. Based on this, the previous equation becomes:

$$AC = FC \cdot q \cdot W + \frac{10 \cdot A}{S \cdot W \cdot E_f} \cdot (RM \cdot q \cdot W + E + k \cdot W + \lambda \cdot W + T_s)$$

According to the corresponding method for the tractor selection, and taking the partial differential referring to the machine size (w), we have:

$$\frac{\partial AC}{\partial W} = FCq - \frac{A \cdot E}{SW^2 E_f} - \frac{A \cdot T_s}{SW^2 E_f}$$

When equating to 0 we have:

$$W = \sqrt{\frac{10 \cdot A \cdot (E + T_S)}{FC \cdot q \cdot S \cdot E_f}}$$

The equation above gives us the machine size with the minimum operating cost.

Example: A farm needs a cultivator for 50 ha per year at a rate of 7 km h^{-1} and 85% yield in the field. Operation cost is 4.5 € l^{-1}. The farm has a tractor with power of 80 kW$_{PTO}$, the fixed costs of which are 9.24 € h^{-1}. The fixed costs of the plow are 13% of the purchase price. The purchase price per meter of processing width is €1500. The size of the plow with the minimum cost is:

$$W = \sqrt{\frac{10 \cdot 50 \cdot (4.5 + 9.24)}{1500 \cdot 0.13 \cdot 7 \cdot 0.85}} = 2.43 \text{ m}$$

The processing cost, with the 80 kW$_{PTO}$ tractor, will be T = 23.16 € h^{-1} for a total annual use of 500 h, RM plow = 0,0004 Q_o h^{-1}:

$$AC = FC \cdot Q_o + \frac{10 \cdot A}{S \cdot W \cdot E_f} \cdot (RM \cdot Q_o + T)$$

Replacing the values:

$$AC = 1500 \cdot 2.43 \cdot 0.13 + \frac{500}{7 \cdot 2.43 \cdot 0.85} \cdot (1500 \cdot 2.43 \cdot 0.0004 \cdot 23.16) = 1640.78 \text{ €}$$

From the above we have:

Cost per ha = €32.8
Cost per hour = €44.65

After calculating the size, the power of the tractor must be calculated so we can see if it is enough.

In this example, the elements in Appendix Table II show that for a cultivator working on average ground on secondary work there is a power requirement of 6.5 kW for a width of 1 m, giving 17 kW in total. Taking into account all the transmission coefficients (TE, P$_{PTO}$ = P$_{A\Xi}$/0.96), power is obtained at PTO = 35.41 kW. The power of the tractor (80 kW) is much higher than required.

Example: The same holding needs a plow for plowing 30 ha per year at a rate of 7 km h^{-1} and a field yield of 80%. Fixed plow costs are 13% of the purchase price. The purchase price per meter of processing width is 1600 € m^{-1}.

The size of plow with the minimum operating cost is:

$$W = \sqrt{\frac{10 \cdot 30 \cdot (4.5 + 9.24)}{1600 \cdot 0.13 \cdot 7 \cdot 0.80}} = 1.88 \text{ m}$$

The plowing cost with an 80 kW tractor and a 1.9 m plow is calculated according to the equation T = tractor cost per hour at a total, EΣ plow: $0.0005\ Q_o\ h^{-1}$:

$$AC = FC \cdot Q_o + \frac{10 \cdot A}{S \cdot W \cdot E_f} \cdot (RM \cdot Q_o + T)$$

By replacing the values:

$$AC = 1600 \cdot 1.88 \cdot 0.13 + \frac{300}{7 \cdot 1.88 \cdot 0.8} \cdot (1600 \cdot 1.88 \cdot 0.0005 \cdot 23.16) = 1383.6\ €$$

From this comes:

Cost per hour = €48.9
Cost per ha = €46.1

The required power, according to the data in Appendix Table I (resistance for average ground and depths of 25 cm = 15,783 N m^{-1} and 30 cm = 18,940 N m^{-1}), will be in the first case 83 kW$_{PTO}$ and in the second 99 kW$_{PTO}$. The power of the tractor is therefore insufficient.

The method has a limited application for high-power accessories, such as plows, heavy cultivators, disk harrows for main processing, soil conditioners, etc., as shown in the second example. The equation gives large sizes because it is assumed that the cost of the tractor (T) is not affected by the size of the component. In reality, however, it is the cost of the tractor and not of the implement that determines the size of the implement.[14] For implements that do not require high power (as in the first example), the results are satisfactory.

In general terms this method calculates the size of implements that actually have the minimum cost but does not estimate whether the work will be done on time. In other words, it defines the size based more on economic terms rather than technical ones.

5.2.4 Calculation of Implement Size Based on Execution of Field Operations on Time

This method, also known as the cost-effective method, is proportional to the corresponding method of the tractor power calculation and takes into account the costs of not carrying out the work on time. The charge factors are the same as those used to calculate the power of the tractor.

Taking into account the problems arising from untimely execution of operations, the operating costs are:

$$AC = FC \cdot q \cdot W + \frac{10 \cdot A}{S \cdot W \cdot E_f} \cdot (RM \cdot Q_o + E + K + L + T_c + M_E)$$

or

$$AC = FC \cdot q \cdot W + \frac{10 \cdot A}{S \cdot W \cdot E_f} \cdot (RM \cdot q \cdot W + E + k \cdot W + \lambda \cdot W + T_c + M_E)$$

where $Q_0 = q \cdot W$, $K = k \cdot W$, $\Lambda = \lambda \cdot W$; $M_E =$ hourly charges due to improper work execution $[(F_c \cdot A \cdot U \cdot V)/Z$ (€ h^{-1})]; $F_c =$ charge factor (per hour); $U =$ crop yield (kg ha^{-1}); $V =$ product sale price (€ kg^{-1}); $Z = 4$ for balanced field operations; $Z = 2$ for early or late field operations; $T_c =$ tractor fixed costs per hour (€ h^{-1}).

The equation below calculates the width of the machine (W):

$$W = \sqrt{\frac{10 \cdot A \cdot (E + T_c + M_E)}{FC \cdot q \cdot S \cdot E_f}}$$

The result of the equation gives a very specific mathematical answer. From a practical point of view, however, the importance of this answer depends on the shape of the annual expenditure curve as a function of the size of the machine. In a curve with a steep dive (very low minimum cost, which varies greatly when the machine width changes slightly), a small change of the size results in a significant change in the annual cost. In contrast, in a smoother curve the variation in width does not have a great effect on the annual cost of operation.

If the producer considers that a default change in the machine's operating cost (e.g., ± 10%) does not play a big role in the cost of the final product, it is possible to choose the size of the implement with the help of the equation[14]:

$$W_{1,2} = W + \frac{d}{2 \cdot FC \cdot q} \pm \sqrt{\frac{d}{FC \cdot q} \cdot \left(W + \frac{d}{4 \cdot FC \cdot q} \right)}$$

where $d =$ default cost change (€).

The importance of the equation is obvious, because it allows more fluctuation when choosing the machine size with a controlled change of cost.

Both the cost-optimal method and the minimum cost method can be used when the implement is used for more crops.

Example: A farm needs to purchase a seed-drilling machine for cotton drilling on 20 ha per year and corn on 5 ha per year. Field speed is 8 km h^{-1} for cotton and 7 km h^{-1} for corn, with field efficiency of 65% and 70%. The yield per area of cotton is 400 kg and the price is 0.80 € kg^{-1}. The yield of corn is 8000 kg ha^{-1} and the price 0.13 € kg^{-1}. The labor costs are 4.5 € h^{-1} and the fixed cost of the tractor is 9.24 € kg^{-1}. The purchase price of the machine is 2500 € m^{-1} and the fixed costs are 13% of the purchase price. The charge coefficients are SE = 0.0003 per hour and the work is done with a balanced plan (Z = 4).

The seed-drilling machine size with optimal operating cost is found with the help of the equation:

$$W^2 = \frac{1}{FC \cdot q} \cdot \left[\frac{10 \cdot A_B}{S_B \cdot E_{fB}} \cdot (E + T_c + M_{EB}) + \frac{10 \cdot A_K}{S_K \cdot E_{fK}} \cdot (E + T_c + M_{EK}) \right]$$

M_{EB}: cost per hour for cotton (3.2 € h^{-1})
M_{EK}: cost per hour for corn (0.73 € h^{-1})
Index B for cotton and index K for corn.
When replacing values:

$$W^2 = \frac{1}{0.13 \cdot 2500} \cdot \left[\frac{10 \cdot 20}{8 \cdot 0.65} \cdot (4.5 + 9.24 + 3.2) + \frac{10 \cdot 5}{7 \cdot 0.7} \cdot (4.5 + 9.24 + 0.73) \right]$$

$$= 2.96 \text{ m}$$

and W = 1.72 m.

The use per year of the machine is 28.29 h.

The value is increased using the cost-effective method to complete the work on time.

The operation cost of running the seed-drilling machine is:

$$AC = FC \cdot Q_o \cdot \left[\frac{10 \cdot A}{S_B \cdot W \cdot E_{fB}} + \frac{10 \cdot A_K}{S_K \cdot W \cdot E_{fK}} \right] \cdot (RM \cdot Q_o + T)$$

Repairs and maintenance costs are 0.0005 Q_o h^{-1}. Total cost of tractor use is 23.16 € h^{-1}.

Replacing the values, we have:

$$AC = 0.13 \cdot 2500 \cdot 1.72$$
$$+ \left(\frac{200}{8 \cdot 1.72 \cdot 0.65} + \frac{50}{7 \cdot 1.72 \cdot 0.7} \right) \cdot (0.0005 \cdot 2500 \cdot 1.72 \cdot 23.16)$$
$$= 1967.8 \text{ €}$$

From this comes:

Cost per ha = €78.7.

If the producer considers that a ±10% change in the annual cost (≈ 200 €) is not significant, the size of the seed-drilling machine is calculated with the equation:

$$W_{1,2} = 1.72 + \frac{10 \cdot 20}{2 \cdot 0.13 \cdot 2500} \pm \sqrt{\frac{10 \cdot 20}{0.13 \cdot 2500} \cdot \left(1.72 + \frac{10 \cdot 20}{4 \cdot 0.13 \cdot 2500} \right)}$$

and $W_1 = 0.87$ m, $W_2 = 3.18$ m.

For a cost change of ±10%, the size of the seed-drilling machine is between 87 and 318 cm.

This method can be used when there is a tractor in the company and its operating costs are known, referring to both the annual total and the fixed costs. However, it should be stressed that the results using this method, as well as the previous one, are not reliable enough when selecting equipment whose resistance to traction is great; but for other machinery it does give reliable results.

5.2.5 Implement Size Calculation Based on Charges Due to Time Limitation and Untimely Field Operations

The calculation of implement size on the basis of both the charges due to untimely field operations and the limitation of time is proportional to the calculation of the required tractor power over a limited period of time.

It differs from the previous method, since hourly charges (costs) due to untimely operation execution are greater because of the limited working hours available. Thus the annual charges due to untimely operation execution are[71]:

$$E_\pi = \frac{F_c \cdot (10 \cdot A)^2 \cdot Y \cdot V}{Z \cdot p_r \cdot C_E}$$

where F_c, U, V, Z as before.AC_c = annual cost of charges due to improper performance of the work (€)C_E = real field efficiency [$S \cdot W \cdot E_f$ (ha h^{-1})] p_Γ = ratio of available days to total calendar days.

The limitation of time, expressed by the ratio of working days to both working and not working days (decimals), results in an increase in the charge due to the untimely operation execution. To use the same example as the previous method, if during the seeding period the working days are 0.5 of the total days, the calculation of the size of the seed-drilling machine can use the same equation as the previous method. The difference is that the charges here are increased. Thus the charges for cotton are 3.2/ 0.5 = 6.4 € h^{-1}, and for corn they are 0.73/0.5 = 1.46 € h^{-1}. The rest of the data remains the same as in the previous method. The equation is therefore:

$$W^2 = \frac{1}{FC \cdot q} \cdot \left[\frac{10 \cdot A_B}{S_B \cdot E_{fB}} \cdot (E + T_\Sigma + M_{EB}) + \frac{10 \cdot A_K}{S_K \cdot E_{fK}} \cdot (E + T_\Sigma + M_{EK}) \right]$$

Replacing the values:

$$W^2 = \frac{1}{0.13 \cdot 2500} \cdot \left[\frac{200}{8 \cdot 0.65} \cdot (4.5 + 9.24 + 6.4) + \frac{50}{7 \cdot 0.7} \cdot (4.5 + 9.24 + 1.46) \right]$$

$$= 2.86 \ m$$

and $W = 1.7$ m.

The annual use of the seed-drilling machine amounts to 28.62 h.

By comparing this value with the one in the previous method we observe a slight increase, which is not significant in any case.

The cost of running the seed-drilling machine, using the same data as the previous example, is:

$$AC = FC \cdot Q_o \cdot \left[\frac{10 \cdot A}{S_B \cdot W \cdot E_{fB}} + \frac{10 \cdot A_K}{S_K \cdot W \cdot E_{fK}} \right] \cdot (RM \cdot Q_o + T)$$

Replacing the values:

$$AC = 0.13 \cdot 2500 \cdot 1.7 + \left(\frac{200}{8 \cdot 1.7 \cdot 0.65} + \frac{50}{7 \cdot 1.7 \cdot 0.7} \right) \cdot (0.0005 \cdot 2500 \cdot 1.7 \cdot 23.16)$$

$$= 1961 \ €$$

From this comes:

Cost per ha = €78.4.

We can see that in comparison with the previous method there is a slight increase in the size of the machine and at the same time a decrease in the annual cost. This results from the reduction in annual use: in the previous case it was 24.66 h, whereas in this case it is 23.12 h.

This way we can also define the minimum and maximum size of the machine, as long as the cost difference in the purchase price is affordable for the farmer. For a cost change of ±10% (≈ 200 €):

$$W_{1,2} = 1.7 + \frac{200}{2 \cdot 0.13 \cdot 2500} \pm \sqrt{\frac{200}{0.13 \cdot 2500} \cdot \left(1.7 + \frac{200}{4 \cdot 0.13 \cdot 2500}\right)}$$

and $W_1 = 0.93$ m, $W_2 = 3.07$ m.

According to the speed, the degree of field yield, the degree of power output in the traction, the working width, and the resistance during the operation of the machine, the power required for traction and operation of the seed-drilling machine is as follows.

W = 3.07 m (maximum)

S = 8 km h^{-1} (maximum cotton)

E_f = 0.65 (minimum cotton)

TE = 0.50

$P_{tr.}$ = 4 km m^{-1} (Appendix Table I).

$$P_{tr.} = 3.07 \cdot 4 = 12.28 \; kW$$

$$TE = \frac{P_{tr.}}{P_{AE}} \text{ and } \frac{P_{AE}}{P_{PTO}} = 0.96$$

$$P_{AE} = \frac{P_{tr.}}{TE} = \frac{12.28}{0.5} = 24.56 \; kW$$

$$P_{PTO} = \frac{P_{AE}}{0.96} = 25.58 \; kW$$

When charged at 75% the power is:

$$P_{PTO} = 25.58/0.75 = 34.1 \; kW$$

The seed-drilling machine requires for its operation a tractor of 34 kW power, thus the 60 kW$_{PTO}$ tractor has more than enough power.

5.3 Machinery Replacement

Farm machines can operate for a long time, usually until the end of their economic life. This depends on many factors, as already mentioned. At the end of this period they should be replaced, because their operation cost gets higher and higher, and can become more than the purchase price of a new machine of the same size.

Farm machinery replacement decisions are very important in farm operation management. A hasty decision to replace due to a temporary machine malfunction or a naive decision to purchase the latest model for reasons of modernization and (perhaps) ego may result in a serious consequential drain of operating capital. On the other hand, delayed replacements are counterproductive. The maintenance function again plays a prominent advisory role in this issue, since it gives the maintenance manager the relevant machine records or data helping her/him to perform her/his role efficiently.

The commonest reasons for equipment replacement are as follows.

- Inadequacy, which results from a change in operating conditions and the consequent incapacity of an existing machine to meet the new requirements.
- Deterioration, which means excessive operating cost, increased maintenance costs, high reject rates, high frequency of stoppages, and increased safety issues.
- Obsolescence arising from technological advancement, which on the one hand makes existing machines less efficient in performance and on the other hand removes spare parts for previous models from the market, making it more and more difficult to find such of these.
- The machine is often immobilized by faults, resulting in reduced reliability and increased charges due to improper work execution.
- As time goes by and the farm changes its productive direction, the old machine is not effective when providing the required services.
- A good selling price for the existing machine is offered.
- Issues of social development and recognition in a close-knit society often contribute to machinery replacement.

A good policy for machinery replacement is very important, since the investment is planned to last for at least a period of 12 years or so. The aims of a good policy are as follows.

- Allow purchases of machine to be planned in advance.
- Ensure that machine fleets are maintained in good order.
- Maximize the use of investment capital.
- Replace machinery on a regular basis over the period.

The decision to replace is usually difficult enough. There is no simple and easy rule that applies to a machine operating under the present complicated working conditions.

The time of replacement generally depends on the cost of use. There are many methods to determine the right moment. The main ones propose the most appropriate time to replace as being when the cumulative repair and maintenance costs are equal to the cumulative depreciation costs; when the cumulative depreciation cost plus the cumulative interest on capital cost plus the cumulative repair/maintenance cost per area unit become minimal[14]; when the cumulative depreciation cost plus the cumulative repair/maintenance cost per unit area become minimal; when the cumulative depreciation cost plus the cumulative interest of capital cost plus the cumulative repair/maintenance cost per operation hour become minimal[3]; when the sum of the machine's initial value plus the cumulative repair and maintenance costs per unit area or per hour of operation becomes minimal; when the cumulative annual net income that the machine produces becomes maximum; or when the average cumulative cost per hour is equal to the annual cost per hour.[73]

The most often used method defines as the most appropriate time for replacement the year that the sum of the cumulative depreciation cost, the interest on capital, and the repair/maintenance cost of the machine per unit reaches its minimum value; this is the method applied below.

The cost analysis showed that a new machine bears very high costs due to the high depreciation and high interest rates on the invested capital. In contrast, the repair and maintenance costs are very limited. Over the years, the cost of depreciation decreases while the cost of repairs rises. At the end of its economic life, the cost of repairs becomes too high. Replacement should occur when the sum of the cumulative depreciation cost, interest cost, and repairs/maintenance cost per unit area becomes minimal. Cumulative cost is summed up over the years of the machine's economic life. For safe results, however, the actual residual value of the machinery, as second-hand machinery, should be assessed and the maintenance and repair costs should be recorded but not valued.

Example: A €35000 grass baler is used for 150 h a year. The residual value is 10% of the initial value. The net interest rate is 6%. The annual inflation rate is 3%. The machine serves 150 ha a year. To find the residual value at the end of each year we use[74] $Q = (1 + i_i)^n \cdot [0.852 - 0.0101 \cdot n^{0.5}]^2$, where n is the year and i_i is the annual inflation rate. To find repair and maintenance costs we use the RM $= Q_o \cdot 0.43 \times (h/1000)^{1.8}$, where h is the total operation hours of the machine. These are applied in Table 33.

As we can see, the cost per ha constantly decreases until the sixth year; from the seventh year it starts to rise again. Thus the appropriate time for the machine replacement is after the sixth year, because the cost increases just after this year. The table shows that the minimum cumulative cost per operation hour occurs when the cumulative cost per unit area is also at its lowest levels. This does not happen when there is a real cumulative recording of operation hours and ha; there is always a little or more deviation concerning the minimum levels. In the table it is assumed that operation per hour is directly linked to surface operation (150 h or 150 ha each year). Note also that in this example the time when cumulative maintenance and repair costs are higher than cumulative depreciation costs (also a replacement criterion) is the same as the time of the minimum cumulative cost per unit area.

Fig. 53 shows schematically the typical change of cumulative cost per ha according to year of economic life.

When replacing machines, a common practice is to replace an old machine with a new one of the same size, as long as the conditions have not changed. The old machines are either sold, usually at their residual value, or remain in the business to be used at any time when the new machinery is immobilized due to a fault. When it comes to tractors, it is common for the old one to remain in the business and be used for undemanding auxiliary operations (transportation etc.). This is due to a number of factors, the main ones of which are the existence of a second

Table 33 Calculation of Appropriate Time to Replace a Grass Baler With an Initial Value of €35000

Year	Cumulative Hours	Cumulative Remaining Value	Cumulative Depreciation	Cumulative Capital Charge	Cumulative Maintenance Costs	Cumulative Cost (Columns 4 + 5 + 6)	Cumulative Use (ha)	Cumulative Cost Per ha	Cumulative Cost Per Hour
1	2	3	4	5	6	7	8	9	10
1	200	20333	14667	1834	831	17331	150	115.5	115.54
2	400	18674	16326	3690	2892	22909	300	76.4	76.36
3	600	17533	17467	5425	6001	28893	450	64.2	64.21
4	800	16644	18356	7037	10072	35465	600	59.1	59.11
5	1000	15909	19091	8526	15050	42668	750	56.9	56.89
6	1200	15276	19724	9894	20896	50513	900	56.1	56.13
7	1400	14720	20280	11138	27578	58997	1050	56.2	56.19
8	1600	14221	20780	12260	35071	68111	1200	56.8	56.76
9	1800	13764	21236	13260	43354	77850	1350	57.7	57.67
10	2000	13343	21657	14138	52407	88202	1500	58.8	58.80
11	2200	12951	22049	14893	62215	99157	1650	60.1	60.10
12	2400	12581	22419	15525	72764	110708	1800	61.5	61.50
13	2600	12233	22768	16035	84040	122843	1950	63.0	63.00
14	2800	11899	23101	16423	96033	135557	2100	64.6	64.55

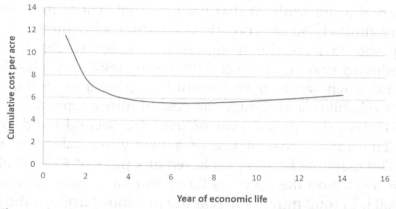

FIGURE 53 Change in cumulative cost per unit area according to years of economic life of a grass-baling machine (minimum at the end of the sixth year).

tractor, although old, helps when undemanding work needs to be done; it also helps when the size of the farm and its needs are increased, often observed in recent years; for emotional reasons; and the sale price is unprofitable or the machine is not easy to sell due to lack of a healthy market for used machinery.

5.3.1 Selection of Second-Hand Machinery

Farm machines lose most of their value in the first years of use, just like cars do. This phenomenon is especially noticed when it comes to tractors and other self-propelled machinery. The purchase of second-hand machines, especially tractors, at an advantageous price after they have been used to only a small extent could be a satisfactory proposition in many cases.

The choice of a second-hand machine has the advantage of low initial capital requirement for the purchase and lower fixed operating costs, due to lower depreciation and interest on invested capital. However, variable costs are expected to be higher due to increased repair and maintenance costs. Compared to new machinery, the reliability of second-hand machines is lower because of their age and frequent breakdowns, which means that some projects may not be able to be carried out on time; thus the indirect cost is high. For these reasons it is preferable to select second-hand machines when their annual use is limited and there is plenty of time to carry out the work, so that the potential cost of improper execution is very limited. Apart from this, a farmer who buys second-hand machinery should have the technical knowledge to repair it herself/himself, reducing cost and time and increasing reliability. Usually when a second-hand machine has been bought it is given a general overhaul, to increase its reliability and reduce the cost of future repairs.

Table 34 gives the annual cost of use of a second-hand grass baler machine. The annual cost of use of a new grass baler was shown in Table 33. The machine here is purchased at a price of €13500 after it has worked for 5 years on the previous farm, and has a new economic life of 1250 h (2000 h in total minus 750 h at the previous farm). At the new farm it works for 100 h per year, serving 100 ha annually.

In Table 33 the depreciation was estimated based on the calculation ratio of the new machine. This leads to an error, because the actual residual values at the end of each year may not be the same as those of the second-hand market, and in times of high inflation the differences may be significant. When there are comparisons with prices of second-hand machines the results are more reliable. The same problems arise concerning maintenance and repair costs. Table 34 was created on the basis of the estimated costs according to the equation followed for a new machine. This leads to rather high expenditure, higher than reality, at the end of the economic life. In practice, for grass balers there is a significant increase of costs for the first 200 h, then a smaller increase for the next 400 h or so of total use, and after this point there is stabilization.[14] Nevertheless, the results of Table 34 represent a satisfactory first approach.

Table 35 shows the comparison of costs between a new and a second-hand machine for the first year of operation.

The comparison of costs between a new and a second-hand machine shows that in the first year, there is a significant cost difference, which justifies the use of a second-hand machine. Smaller differences exist in the following years as well.

Despite the economic benefits offered by market for second-hand machines, there is not much preference for them. The main reasons for this are firstly a fear that the second-hand vehicle is not in good condition, so a lot of money will needed to repair it. When a farmer decides to buy a

Table 34 Cost of Use of Second-Hand Grass Baler After 5 Years of Previous Use (750 h)

Year	Cumulative Hours	Remaining Value	Cumulative Depreciation	Cumulative Capital Charge	Cumulative Maintenance Costs	Cumulative Cost (Columns 4 + 5 + 6)	Cumulative Use (ha)	Cumulative Cost Per ha
1	2	3	4	5	6	7	8	9
1	100	850	541	798	1942	3281	100	32.8
2	200	950	1018	1569	4076	6664	200	33.3
3	300	1050	1447	2313	6398	10158	300	33.9
4	400	1150	1838	3033	8904	13775	400	34.4
5	500	1250	2199	3730	11590	17519	500	35.0
6	600	1350	2535	4406	14455	21396	600	35.7
7	700	1450	2851	5063	17494	25408	700	36.3
8	800	1550	3151	5701	20705	29557	800	36.9
9	900	1650	3436	6321	24087	33845	900	37.6
10	1000	1750	3710	6925	27637	38272	1000	38.3
11	1100	1850	3973	7513	31353	42839	1100	38.9
12	1200	1950	4228	8085	35233	47546	1200	39.6

Table 35   Compare Operation Costs of New and Second-Hand Grass Balers

Cost Elements	Grass Baler	
	New	Used
Purchase price (€)	30000	13500
Depreciation (€)	12572	541
Interest (€)	1423	798
Repairs/maintenance (€)	424	1942
Total (€)	14419	3281
Cost/ha (€ ha^{-1})	96.1	32.8
Cost/ha (€ h^{-1})	96.13	32.81

second-hand machine s/he usually prefers to buy from a person s/he knows, so s/he knows the way the machine has been used, or from big companies that import and trade agricultural machinery. In recent years these companies sell second-hand machinery that they have obtained either through replacements or seizures, and then checked, repaired, and guaranteed for their proper operation. Secondly, in the second-hand market there are not always machines of the type and size needed, so the farmer is either forced to buy anything that is available or will choose to buy a new machine. Thirdly, often social reasons of self-promotion in the closed community of a village or small town, as mentioned before, discourage the market for second-hand machines.

In conclusion, the purchase and use of second-hand machinery seem to be a satisfactory solution under certain circumstances, and this possibility should be assessed before each purchase decision.

5.4 Machinery Management System Selection

The term "machinery management systems" refers to types of machine usage according to their ownership status. An agricultural machine or vehicle may be self-owned, under contract, shared by more than one farmer, etc. The selection of the system depends on a number of factors, including the size of the holding, the ownership status of the holding (self-owned, shared, a cooperative, a corporation, or even a state-owned holding), the type of crops, the weather conditions of the area, and the holding's finances. In small holdings farmers usually use self-owned machines and equipment for basic field operations, while for operations requiring expensive or specialized machines, high-power tractors, or machines scheduled to operate for a very short period of time they use the solution of contracting.

The most common machinery management systems are the use of proprietary machinery, the use of professional machinery, cooperative use of machinery, tenancy (leasing) and the use of state-owned machinery.[66]

5.4.1 The Use of Self-Owned Machinery

The use of self-owned machines is the usual management form in all countries with a free market. In its most common form the farmer has one or more agricultural tractors with the appropriate implements. In a very few cases a farmer buys large (self-propelled or not) harvesting machinery for business use.

The main reasons why farmers prefer self-owned machinery are that the farmer feels independent and can complete the appropriate operations in time if the size of the implement is sufficient; the operation cost of self-owned machinery is usually lower, taking into account the cost of untimely operations, and self-owned machinery allows the farmer to operate on other holdings using her/his machines to increase her/his income.

Although there are serious incentives for producers to use their own machines, it should be noted that they require considerable capital which may sometimes not be available.

If a machine only operates for a short time due either to its large size or to reduced needs, the operation cost is high and economically it will be more beneficial for the farmer to choose another ownership system.

Example: A farm buys a 100 kW tractor for €90000 with the intention to rent it to other holdings, on a business basis, for deep harrows. The annual fixed costs amount to €7010. The farm's needs can be covered with a 70 kW tractor worth €60000, which will operate 500 h at an annual cost of €11325, of which €5500 are fixed costs. The power of the bigger tractor allows plowing to a depth of 30 cm and a width of 105 cm (three plows of 35 cm), with operation speed 7 km h^{-1}, TE $= 0.65$ and load at 80% in medium-heavy soil. Its annual use on the holding is estimated at 410 h, slightly less than that required for the 70 kW tractor. The farmer intends to rent it for plowing on other farms. The real performance on the field is CE $=$ S$\cdot$W$\cdot$0.8 $= 7 \cdot 1.05 \cdot 0.8 = 0.588$ ha h^{-1}. The fee when operating in other holdings is 80 € ha^{-1}. The hourly gross income coming from this operation is therefore 0.588 (ha h^{-1}) $\cdot$ 80 (€ ha^{-1}) $= 47.04$ € h^{-1}. According to this data the tractor needs annual operation time in other holdings that keeps the cost at least at equal levels, if not lower, compared to the cost of the smaller tractor that is actually required. Additional information is given: RMC $= 0.000083$ Q_o h^{-1}, E $= 4.5$ € h^{-1} for both tractors, $F_1 + L_1 = 4.947$ € h^{-1} of the tractor of 70 kW, and $F_2 + L_2 = 9.894$ € h^{-1} of the tractor of 100 kW.

If AC_1 and AC_2 are the annual cost of a 70 kW and a 100 kW tractor respectively and AF is the annual fee from the operation in other holdings, we have:

$$AC = AC_2 - AF$$

where

$$AC_1 = FC \cdot Q_1 + t \cdot (RMC \cdot Q_1 + E + F_1 + L_1)$$

$$AC_2 = FC \cdot Q_2 + (t_1 + t_2) \cdot (RMC \cdot Q_2 + E + F_2 + L_2)$$

$$AM = t_2 \cdot V$$

and $t = 500$ h, the operation time of the 70 kW tractor; $t_1 = 410$ h, the operation time of the 100 kW tractor in the holding; $t_2 =$ asked operation time of the 100 kW tractor in other holdings; $V =$ payment from operations in other holdings (€ h^{-1}).

The above equation can be written as:

$$C_1 + t \cdot C_2 = C_3 + (t_1 + t_2) \cdot C_4 - t_2 \cdot C_5$$

$$t_2 = \frac{C_3 + t_1 \cdot C_4 - C_1 - t \cdot C_2}{C_5 - C_4} \quad (h)$$

where $C_1 = FC \cdot Q_1$ (70 kW tractor) (5500 €); $C_2 = FC \cdot Q_1 + E + K_1 + L_1$ (70 kW tractor) (11.65 € h^{-1}); $C_3 = FC \cdot Q_2$ (100 kW tractor) (7010 €);

$C_4 = FC \cdot Q_2 + E + K_2 + L_2$ (100 kW tractor) (17.10 € h^{-1}); $C_5 = V$ (payment from foreign work) (47.04 € h^{-1}).

With value replacements:

$$t_2 = \frac{7010 + 410 \cdot 17{,}1 - 5500 - 500 \cdot 11.65}{47.04 - 17.10} = 90.05 \; h$$

$t_2 = 90.05$ h or $A = 52.95$ ha.

For the 100 kW tractor to have the same annual cost as the 70 kW tractor, it must operate in other holdings for 90.05 h or 52.95 ha. So the 100 kW tractor will work 410 h in the farmer's holding and 90.05 h in other holdings, to give a total of 500.05 h.

In an extreme case where the fee for the work is equal to the total cost of use, the coefficients C_1 and C_2 are equal to zero and the calculation of the time t_2 is:

$$t_2 = \frac{C_3 + t_1 \cdot C_4}{C_5 - C_4} \; (h)$$

$$t_2 = \frac{7010 + 410 \cdot 17.10}{47.04 - 17.10} = 466.7 \; h$$

According to this, the tractor must work in other holdings for 466.7 h or 274.42 ha, giving 876.7 h in total. The same result we have when the cost of use of the 100 kW tractor for operations in and out of the holding is equal with the fee received from the plowing out of the holding:

$$C_3 + (t_1 + t_2) \cdot C_4 = t_2 \cdot C_5$$

5.4.2 Use of Professional Machinery

The term "professional machinery" refers to machinery that the farmer can rent, including the machine operator, to carry out specific work in other holdings at an agreed fee. The cost of professional machinery use is stable per surface unit or time, whereas the use cost of self-owned machinery takes the form of a curve with a minimum.

The breakeven where the use cost of self-owned machines is equal to the cost of professional machines is calculated by the equation:

$$AC = FC \cdot Q_o + \frac{10 \cdot A}{S \cdot W \cdot E_f} \cdot (RMC \cdot Q_o + E + K + L + T)$$

Or, in a more simple equation:

$$AC = C_1 + A \cdot C_2$$

where $C_1 = FC \cdot Q_o$.

$$C_2 = (RMC \cdot Q_o + E + K + L + T)/ S \cdot W \cdot E_f$$

The rental of the professional machinery is stable per surface unit (€ ha^{-1}), so it can be written as:

$$AF = C_3 \cdot A$$

Therefore, when the usage cost of self-owned machinery becomes equal with the cost (rental) of professional machinery, we have:

$$C_1 + A \cdot C_2 = A \cdot C_3$$

where

$$A = \frac{C_1}{C_3 - C_2} = \frac{Fixed\ costs\ (€)}{Rental\ (€\ acre^{-1}) - Variable\ costs\ (€\ acre^{-1})}$$

The above equation expresses the breakeven in ha. However, the rental can be calculated per hour (€ h^{-1}), in which case the variable costs of the self-owned machine should be calculated per hour as well. So the annual cost equation is:

$$AC = FC \cdot Q_o + t \cdot (RMC \cdot Q_o + E + K + L + T)$$

$$AC = C_1 + t \cdot C_2$$

where $C_1 = FC \cdot Q_o$

$$C_2 = (RMC \cdot Q_o + E + K + L + T)$$

The rental (fee) of the professional machinery is stable per time unit (€ h^{-1}), so it can be written as:

$$AF = t \cdot C_3$$

Therefore, when the use cost of the self-owned machinery is equal to the cost (rental) of the professional machinery, we will have:

$$C_1 + t \cdot C_2 = t \cdot C_3$$

$$t = \frac{C_1}{C_3 - C_2} = \frac{Fixed\ costs\ (€)}{Rental\ (€\ acre^{-1}) - Variable\ costs\ (€\ acre^{-1})}$$

The above equation gives the break even in hours.

Example: The fee for seeding 5 ha of grain with a professional tractor/seed-drilling machine is €50 per ha. If the price of the seed-drilling machine is €4200, the fixed costs are 13% of the price and the variable costs are 31.45 € h^{-1}, the operation speed is 8 km h^{-1}, the seeding width is 2 m, and the efficiency on the field is 0.7, the breakeven (in hours and acres) is:

Fee is €50 ha^{-1}

CE = $(8 \cdot 2 \cdot 0.7) = 1.12$ ha h^{-1}

Fee is $5 \cdot 11.2 = 56$ € h^{-1}

FC = €546

Variable costs = 31.45 € h^{-1}

Variable costs = 28.1 € ha^{-1}

$$\text{Break even (hours)} = \frac{546}{56 - 31.45} = 22.24(h)$$

$$\text{Break even } (10 \cdot \text{ha}) = \frac{546}{5 - 2.81} = 249.3(h)$$

Figs. 54 and 55 present the cost of seeding with self-owned and professional machinery and the breakeven according to the area and the operation hours.

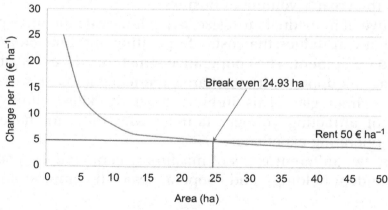

FIGURE 54 Cost of seeding 5 ha of grain with self-owned and professional machinery and breakeven area.

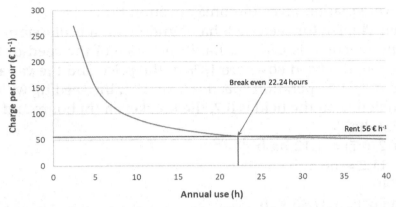

FIGURE 55 Cost of seeding 5 ha of grain with self-owned and professional machinery and breakeven hours of use per year.

5.4.3 Cooperative Use of Machinery

The term "cooperative use of machinery" refers to the exploitation of one or more machines jointly by more than one farmer. Cooperative use may enable farmers to take advantage of technological changes but with lower financial commitments. Thus the advantage of machinery cooperatives is mainly access to newer and more efficient equipment at a lower individual cost. Machinery cooperatives can benefit from price discounts on purchased or leased equipment and operating inputs such as fuel simply because of the greater volume of business.

Irrespective of individual farm size, a machinery cooperative provides a structure that can reduce the costs of operating and financing farm machinery and equipment. The cooperative structure addresses labor issues by providing the opportunity to share experience and skills, particularly with new technologies. This indirect form of skills training has the advantage of attracting younger farmers who may not have a lot of machinery operating experience.

There are two different types of machinery cooperatives: cooperatives with independent holdings, and cooperatives with common holdings.

a) Cooperatives with independent holdings.

In this cooperative type the member of the cooperative has her/his own independent farm holding in which s/he works herself/himself with her/his family. In this type of cooperative the farmer decides on the type of crop to cultivate. Usually s/he has her/his own tractor and agricultural implements. In most cases specialized harvesting machinery (corn combines, sugar beet machines, potato harvesters, etc.) are bought from the cooperative and used on the holdings of a farmer who is a member of the cooperative. The capital for the purchase of the machinery is available from the cooperative via a loan or the cooperative member's fee. Usually a cooperative member operates the machinery and not the farmer. The operation cost of the machinery is allocated proportionately to all the cooperative members.

The basic advantage of these cooperatives is the use of machinery at the lowest cost possible.

b) Cooperatives with common holdings.

In this type of cooperative, members voluntarily contribute part or all of their holding and machinery to the cooperative to create a united holding.

5.4.3.1 *Agricultural Machinery Rings*

A more advanced form of management in groups is the agricultural machinery ring found in countries in Northern Europe, such as Germany, France, the Netherlands, the Scandinavian countries, and the United Kingdom. The groups include producers, professionals, and a manager.[3] There are periods when some partners need a service, others provide this service, and possibly there are professionals who provide the service as well. Some partners may offer the service to themselves. The manager has the main task of organizing the work so that it is done on time. The partners pay a small amount to the manager, depending on the size of the holding. The machines are operated by their owners. The group carries out operations on a holding when a farmer is unable to do so for some reason (illness, accident, working off the farm, etc.). The income generated by the services offered among the partners is invested in the purchase of new machinery, so the group always has reliable machines.

Except for the implements, large harvesting machines and other large self-propelled machines could be used on a cooperative or a group basis since their purchase is very expensive and the period of their usage limited. This way such large machines are used more effectively, resulting in lower operating costs on the one hand and higher income for the producer on the other.

In conclusion, cooperative or group management of machines can be of important help to small farms as long as the procedure is well organized. Despite the problems of this kind of management system, producers can be well equipped at low cost, and earn higher income. Group use is often promoted by the EU, which provide subsidies for the purchase of machinery.

5.4.4 Cooperative Machinery Use

As an alternative to the individual farmer purchase of farm machinery a cooperative may enable the farmers to take advantage of technological changes with lower financial commitments. The advantages of machinery cooperatives come from cost savings associated with access to newer and more efficient equipment at a lower individual cost due to economies of scale. Machinery cooperatives can benefit from price discounts on purchased or leased equipment and purchasing operating inputs such as fuel simply because of the greater volume of business.

Irrespective of individual farm size, the machinery cooperative provides a structure that can reduce the costs of operating and financing farm machinery and equipment. The cooperative structure addresses labor issues by providing the opportunity to share experience and skills, particularly with new technologies. This indirect form of skills training has the advantage of attracting younger farmers who may not have years of machinery operating experience.

5.4.5 Renting/Leasing of Machinery

Another two management systems for agricultural machinery are renting and leasing.

When a farmer rents machinery, a written agreement is signed between her/him and the machine owner that allows the farmer to use it for a certain period of time at a certain fee (rent). The farmer operates the machine and bears the cost of fuel and lubricants. Repairs and maintenance costs are borne by the owner of the machine. In case of damage, the machine is replaced. The duration of the rental may be short (1 day–1 week) or very long (2–3 years).

Leasing differs from renting. In leasing the farmer is responsible for repairs and maintenance, and in the event of damage the machine is not replaced. The leasing usually lasts for 2–3 years or less (2–3 months). A leasing fee is lower than a rental fee. At the end of the leasing period the farmer has the opportunity to buy the machine at a preset price.

The main advantages of renting are that no large capital is required for investment in machinery; if there is an organized market for renting, the farmer has the opportunity to use expensive machines for heavy or special work (large tractors, etc.) for as long as needed, but with low operating costs; and in countries where income from farming is taxed, rental costs are deductible from taxable income.

Renting a machine offers great advantages when the machine will be used intensively. Rents are usually a percentage of the machine's value. For 1 day to 1 week the rent ranges from 1% to 5% of the machine's value, whereas for 2–3 months it ranges between 20% and 30%.[14]

The advantages of leasing are roughly the same as those of renting. An extra advantage is that the farmer may purchase the machine at the end of the leasing period although s/he may already have deducted the leasing fee from her/his taxable income .

5.4.6 Use of State Machinery

In this system, the machinery belongs to the state and is operated by state officials. Repairs and maintenance take place in state-owned garages, usually located in the machinery area. The machines may operate on state or private farms. In the first case, all the costs are borne by the service managed by the machinery, i.e., by the state. When operating on private farms, part of the costs is borne by the farmer, usually a low or even symbolic price paid to the machinery management department.

In guided economies, this system is dominant in the form of state machinery cooperatives on cooperatives where land belongs to the state. The system seems to be able to deliver, under particular circumstances, to meet the needs of agriculture.

In these cases, it is the social benefit that is important rather than the cost of use, which in most cases is very high. Typical factors that contribute to high costs include, but are not limited to, the choice of an unsuitable size, low annual use, poor maintenance, lack of diligent care, indifference by the employee–operator of the machine, general malfunctioning of the state apparatus, etc.

6

Operations Management

Based on the ASABE Standards (American Society of Agricultural and Biological Engineers), agricultural operations management involves four phases.

- Planning: this phase defines the objective of the system, selects system components, and predicts the expected system performance.
- Scheduling: the time when each operation must be performed in the system is determined by taking into account time available, availability of labor, the priorities of the various tasks involved, and the crop requirements as the most important factors in such a decision.
- Operating: this phase concerns the actual execution of operations.
- Controlling: this phase concerns decisions that are made during operations' execution to control the system under the defined productivity measures and standards.

Operations management of a system is separated into different planning levels; five such levels are commonly defined in engineering management science.[5]

- Strategic: decisions at the strategic level concern the design of the production system. For agricultural production this covers 1–5 years or a period that corresponds to at least two cropping cycles. The labor and machinery are considered alongside the crop types.
- Tactical: for the selected cropping plan at the strategic planning level, resource usage is adjusted at this level for a shorter time period of 1–2 years or one to two cropping cycles.
- Operational: this level concerns decision-making in the current cropping cycle, including short-term determination of activities, their timing, and formulation of the jobs and tasks involved.
- Execution: this level corresponds to the controlling phase, and concerns control of the executed jobs and tasks.
- Evaluation: this level concerns comparison of planned and actual executed jobs and tasks based on predefined measures of performance.

Operations Management in Agriculture. https://doi.org/10.1016/B978-0-12-809786-1.00006-0
159

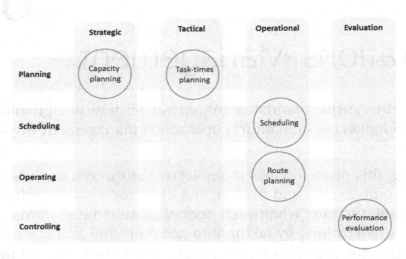

FIGURE 56 The management phases (vertically), the management levels (horizontally), and the management tasks within operations management in agriculture.[91]

Figure 56 presents the interconnections between the management phases and the management levels mentioned above.

6.1 Optimization

All management tasks involve the implementation of an optimization, that is the generation of an optimization problem and the process of solving it. An optimization problem involves finding the values for the set of parameters which define a strategy that returns the best performance of the system. The general formulation of an optimization problem is:

$$minimize \quad J(x)$$

$$subject\ to \quad g_i(x) \le b_i \quad i = 1.....m$$

where vector $x = (x_1.....x_n)$ is the optimization variable of the problem, the function $J : R^n \to R$ is the objective function[a] to be minimized over the

[a]In the case of multiobjective optimization problems, the objective function is given by:

$$J : X \to R^k,\ J(x) = (f_1(x), \dots , f_k(x))^T,$$

where X is the feasible set of solutions.

vector x, $g_i(x)$ are the inequality constraints of the problem, and the constants $b_1.....b_m$ are the bounds for the constraints. The optimal solution of the problem is the vector $x^* = argmax\, J(x)$ that has the smallest objective value among all vectors that satisfy the condition:

$$\forall\, z \ni g_1(z) \le b_1......g_m(z) \le b_m \rightarrow J(z) \ge J(x^*)$$

When solving a problem using simulation, the vector $x = (x_1.....x_n)$ of the strategy parameters is taken as input, and the output returns the value of the objective function $J(x)$. In the case of the presence of unknown or random variables, common in agricultural production optimization problems due to their dependence on stochastic parameters, e.g., weather conditions, then the optimal solution is written as:

$$x^* = argmax\, E[J(x.\xi)]$$

The simulation solution is usually estimated by averaging the objective function over a large number of stochastic parameters, ξ_i. $i \in \{1.....M\}$:

$$E[J(x.\xi)] \approx \frac{1}{M} \sum_{i=1}^{M} J(x.\xi_i)$$

6.2 Capacity Planning

Capacity planning is defined as "the qualitative and quantitative selection of components of a production system as related to the demand" and is part of the system design process.[91]

As illustrated in Fig. 57, the capacity planning process is governed by the operational demands, equipment available, working methods, the capacity and dimensions of the available resources (taking into account the effects of timeliness and workability restrictions), and finally the cost. The main task of capacity planning in agricultural operations, analogous to industrial systems, is optimal allocation of the system's components both in time and in terms of components' operational interaction. However, in agriculture planning is governed by the uncertain parameters of biological and weather conditions.

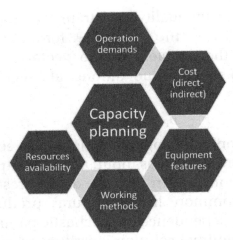

FIGURE 57 Entities governing capacity planning process.

6.3 Task Time Planning

6.3.1 Scheduling

The problem of allocating finite resources over time with various constraints arises in many application fields, such as project management, manufacturing, computing and networking, human resources management, logistics, robotics, and many others. One of the most common problems in this area is project scheduling. Project scheduling problems[92] (PSPs) are made up of activities, resources, precedence relations, and performance measures. A project consists of a number of activities, also known as jobs, operations, or tasks. To complete the project successfully, each activity has to be processed in one of several modes. Each mode represents a distinct way of performing the activity under consideration. The mode determines the duration of the activity, measured in number of periods, which indicates the time taken to complete the activity, the requirements for resources, and possible cash inflows or outflows occurring at the start, during processing, or on completion of the activity. The presence of modes complicates the PSP by requiring assignments of activities to different modes. Often technological reasons mean that some activities have to be finished before others can start. This is handled by depicting the project as a directed graph where an activity is represented by a node and the precedence relation between two activities is represented by a directed arc. Also, minimum and maximum time lags may be given between the finish of an activity and the start of the next one.

Resources utilized by the activities are classified according to categories, types, and value. The category classification includes resources that are renewable, nonrenewable, partially renewable, or doubly constrained. Each resource type has a value associated with it, representing the available amount. Whenever there is at least one category of constrained resources the problem becomes a resource-constrained PSP. Makespan minimization is probably the most researched and widely applied objective in the project scheduling domain. The makespan is defined as the time span between the start and the end of the project. Since the start of the project is usually assumed to be at $t = 0$, minimizing the makespan minimizes the maximum finish times of all activities. Other performance measures are minimization of the (weighted) flow time of the activities or, if due dates are given, minimization of the (weighted) delays, activity plus resource cost, and others.

In the standard PSP the duration or processing of an operation is fixed and known. A generalization of this setting that is still deterministic (and assumes complete information) is obtained by permitting processing times to vary according to how much the planner is willing to pay. This control on processing times can be interpreted as allocation of a nonrenewable resource to the activities, where a larger allocation to an activity (i.e., a higher cost input) reduces its processing time. As the allocation is usually measured in money, these are commonly known as time–cost trade-off problems.

In reality the processing time of activities may not be fixed because it is subject to unpredictable changes due to unforeseen events (weather conditions, obstruction of resource use, delay of predecessors of an activity, etc.). To cope with such influences, the processing time of an activity is assumed to be a random variable. This is known as stochastic project scheduling, and leads to the area of stochastic dynamic programming. For complexity and stability reasons, scheduling is done by relatively simple policies or strategies.[93–95] The complexity of such problems is open, but there is evidence that is also NP (Non-deterministic Polynomial time).

6.3.2   Job-Shop Scheduling

The standard deterministic job-shop scheduling problem (JSSP) involves assignment of a set of jobs to machines in a predefined sequence to optimize one or more objectives considering job performances measures of the system. A job-shop environment consists of n jobs and m machines, and each job has a given machine route in which some machines can be missed and some can repeat. Each job requires a fixed, known processing time on any given machine. The goal is to assign jobs to specific machines according to a schedule which minimizes criteria like the makespan (end of last operation), the average tardiness, etc. The number of all possible solutions of the standard JSSP is in the order of $(n!)^m$ and cannot be exhaustively enumerated even for moderate-sized problems. Unfortunately, the standard JSSP is known to be NP-complete when $m \geq 3$, which means that no nonenumerative algorithms are known.[96] The deterministic JSSP has been studied extensively because of its importance for manufacturing, and numerous approaches have been used to solve it.[97]

The standard deterministic JSSP has been extended to problems in which the execution time for a job is increased by a setup time, which is dependent on the sequence of previous jobs in the machine. For example, a certain job after completion may impose more cleaning and retooling of the machine than another one before the next job is serviced. This problem is equivalent to the multiple traveling salesman problem (TSP), and is even harder to solve than the standard JSSP. Another variation is the dynamic JSSP, in which the number of jobs is not fixed and new jobs may arrive at any time according to some probability distribution. Furthermore, when job setup or processing times are not fixed but random, the problem becomes one of stochastic job-shop scheduling. Most research in production scheduling is concerned with the minimization of a single criterion. JSSP has been recently extended to the multiobjective vehicle routing problem (VRP).[98] A multiobjective optimization problem is one in which two or more objectives or parameters contribute to the overall result. These objectives often affect one another in complex, nonlinear ways. The challenge is to find a set of values for them which yields an optimization of the overall problem. Instead of an optimal solution, a set of solutions is computed which form the Pareto optimal of different objectives.

If agricultural machines are thought of as manufacturing machines in a machine shop and field operations as jobs, then the fleet management problem (FMP) looks similar to some variant of the deterministic JSSP, in particular to one with sequence-dependent setup time. This is because in the FMP the assignment cost at time t_k depends on the current state of the machines of a candidate assignment and consequently on the previous operations they were handling. For example, assigning a machine to work in field A will cost less if the machine is already operating there than if it is currently working in distant field B. The FMP cannot be expressed trivially as a JSSP because an agricultural field operation can be performed by more than one machine in parallel; this is not the case for JSSP, where a major assumption is that each operation is performed by only one machine. The FMP involves machine allocation (coalition formation) in conjunction with scheduling.

6.3.3   Solution Techniques

Moderate and large resource scheduling problems are too difficult to be solved exactly within a reasonable amount of time, and *heuristics* become the methods of choice. Common heuristics are those based on applying some sort of greediness or priority-based procedures, including, e.g., insertion and dispatching rules. As an extension of these, a large number of local search approaches has been developed to improve given feasible solutions. The main drawback of these approaches is their inability to continue the search upon becoming trapped in a local optimum, and this leads to consideration of techniques for guiding known heuristics to overcome local optimality. Intelligent search methods like the tabu search constitute *metaheuristics* for solving optimization problems. Other metaheuristics include evolutionary algorithms, simulated annealing, and ant-colony optimization. Many state-of-the-art metaheuristic developments are too problem-specific or knowledge-intensive to be implemented in cheap, easy-to-use computer systems.

There is a vast literature available on the solution of PSP problems.[92,99] Exact procedures offer globally optimal solutions for small problems, whereas heuristic procedures may solve larger problems without guarantees of optimality. Exact approaches include dynamic programming, zero—one programming, and implicit enumeration with branch-and-bound. Heuristic techniques include single pass and multipass priority rule-based scheduling and truncated branch-and-bound, and metaheuristics include genetic algorithms. A promising approach is the constraint propagation technique, in which model-based local reasoning over the constraint set makes problem-specific knowledge, which is implicitly contained in the model description, explicitly available. The goal is to accelerate exact algorithms or local search procedures.

JSSP is one of the hardest combinatorial optimization problems. A large number of different solution approaches has been developed over the years. Exact methods use integer programming formulations and enumeration techniques like branch-and-bound[99] to find a globally optimal solution. Exact methods can only solve relatively small problems. Many approximate (meta)heuristic methods have also been proposed which do not guarantee globally optimal solutions but can tackle problems of large size. These include the shifting bottleneck procedure,[100] simulated annealing,[101] tabu search,[102] genetic algorithms,[103] ant algorithms,[104] artificial immune systems,[105] and constraint programming.[106]

VRP constitutes one of the most challenging combinatorial optimization problems, and a large number of different problem-solving approaches have been developed over the years.[107–109] Exact methods use integer programming formulations and techniques like branch-and-bound, branch-and-cut, and branch-and-cut with pricing to find a globally optimal solution, but can only solve relatively small problems. Many inexact heuristic methods have also been proposed which do not guarantee globally optimal solutions but can tackle problems of large size. These include simulated annealing,[110] tabu search,[111] genetic algorithms, ant algorithms,[112] and constraint programming.[113]

As can be seen, all problems share similar exact or approximate solution methods, based on heuristics. Good heuristics are extremely important because they enable the solution of very large problems. Heuristics depend on the application domain and require insight, hence they are hard to invent. An alternative approach is learning appropriate heuristics. For example, Zhang and Dietterich presented a reinforcement learning approach to scheduling that learned domain-specific heuristics for the scheduling procedure.[114] The state space consisted of possible schedules, and actions were possible changes to the schedules.

The system learned what changes would quickly create feasible schedules with maximized capacity utilization. The problem domain considered was space shuttle payload processing. Zomaya et al.presented another algorithm for learning scheduling heuristics[115] which learned dynamic scheduling, i.e., scheduling when there is no a priori knowledge about the tasks. It used a back-propagation neural network and a history queue that functions like an eligibility trace to learn how to associate a set of job parameters with a set of machines.

6.3.4   Scheduling Problem Types in Agricultural Operations

The type of scheduling problem in field operations is determined by the units involved in an operation (e.g., machines and machinery features), the particular constraints of the operation, and the objective of the problem (for example time or total cost).[116] A classification based on these elements is presented in Fig. 58.

Regarding units, scheduling operations always include multiple-machinery systems. These systems can operate, depending on the main function, either in parallel or in series. When units operate in parallel they can be either identical or nonidentical in terms of their capacity. Example for identical units is the execution of operations of machines of the same type and characteristics in multiple fields, still however, a theatrical scenario that applies in the case of robotics under the hypothesis of the replacement of large conventional units by smaller autonomous ones. The difference that characterizes two units as nonidentical is the capacity, where the term capacity refers to the ability to perform. In the case where the capacity of each unit does not depend on the field features on where the unit operates, then the above cases are generalized to the case of unrelated units operating in parallel.

The commonest case of machines operating in parallel belongs to a well-known problem in industry, flow-shop scheduling. In the flow-shop scheduling problem in its general form, n jobs require operations on m machines in the same sequence for a given processing time for each operation p_{ij} where $i \in \{1.....n\}$ denotes a job and $j \in \{1.....m\}$ denotes a machine. The problem is to determine the permutation of the n jobs that results in the smallest makespan without preemption of operations. For agricultural operations this involves the allocation of fields to machines

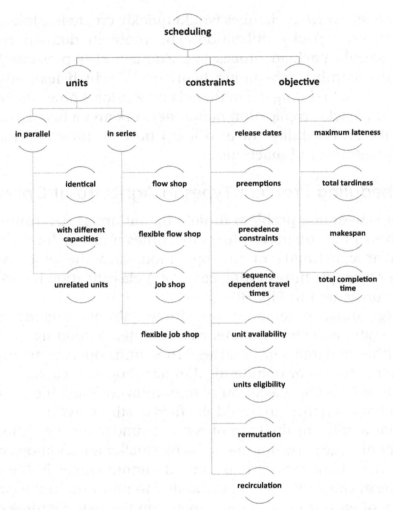

FIGURE 58 Classification of field operation scheduling problems.

where the operations in all fields have to follow the same sequence. This type of problem involves an order discipline, as for example the first-in first-out discipline (permutation flow shop).

Other types of relative problems are the flexible flow shop, where the operation involves units that operate both in parallel and in series, the job shop, where each field has its own predetermined sequence of operations that have to be carried out, and the flexible job shop, a generalization of the job shop involving units operating in parallel and in series.

6.4 Agricultural Vehicle Routing

The majority of route planning cases for agricultural vehicles belong to the well-known VRP family of optimization problems. VRP, in general, is an optimization problem in which the solution determines the routes with the minimum cost (according to one or more criteria) that connect the initial position (depot) of a vehicle to a set of geographically dispersed points which can be, depending of the physical problem, a number of cities, warehouses, schools, etc. The VRP is a generalization of the TSP,[117] which is periphrastically defined as given a set of cities, an origin city among them, and the distances between each pair of cities, find the shortest path that visits each city exactly once and returns to the origin city. Formally the TSP is defined as for a set of cities (or nodes. or points) $V = \{1.....n\}$, the edge set $A = \{(i.j): i.j \in V\}$, and a cost measure $c_{ij}. (i.j) \in A$ associated with the edge set, find a closed tour with minimal length that visits all cities of V exactly once. In a case where $c_{ij} = c_{ji}$ the problem is symmetric TSP (sTSP), while in a case where $c_{ij} \neq c_{ji}$ the problem is called asymmetric TSP (aTSP). Using graph theory terminology, the aTSP can be defined on a directed graph[b] $G = (V.E)$ where $E = \{(i.j): i.j \in V; i < j\}$, while the aTSP can be defined on an undirected graph[c] $G = (V.A)$ where $A = \{(i.j): i.j \in V; i \neq j\}$ and a cost matrix $C = (c_{ij}). (i.j) \in E$ *or* A, respectively. For the mathematical formulation of the TSP as an optimization problem, a binary decision variable $x_{ij}. (i.j) \in E$ *or* A for cases of an undirected or directed problem, respectively, is defined, where: $x_{ij} = 1$ if the edge $(i.j)$ belongs in the tour and $x_{ij} = 0$ if the edge $(i.j)$ does not belong in the tour.

Based on these definitions, the aTSP can be formulated (among other formulations[47,48]) as an integer programming problem:

$$min \sum_{(i.j) \in E} x_{ij} c_{ij}$$

Subject to:

$$\sum_{i<k} x_{ik} + \sum_{k<j} x_{kj} = 2 \qquad k \in V \tag{1}$$

[b]A graph in which edges have no orientation.
[c]A graph in which edges have orientations.

$$\sum_{i.j\in S} x_{ij} \le |S| - 1 \quad (S \subset V.\, 3 \le |S| \le n-3) \tag{2}$$

$$x_{ij} = 1 \; or \; 0 \quad (i.j) \in E \tag{3}$$

Constraints (1) are the degree constraints; constraints (2) are the sub-tour elimination constraints; and constraints (3) refer to integrality constraints.

The aTSP can also be formulated as another integer programming problem[120]:

$$min \sum_{(i.j)\in A} x_{ij} c_{ij}$$

Subject to:

$$\sum_{i=1}^{n} x_{ij} = 1 \quad (j \in V.\, j \ne i)$$

$$\sum_{j=1}^{n} x_{ij} = 1 \quad (i \in V.\, j \ne i)$$

$$\sum_{i.j\in S} x_{ij} \le |S| - 1 \quad (S \subset V.\, 2 \le |S| \le n-2)$$

$$x_{ij} = 1 \; or \; 0 \quad (i.j) \in E$$

The next generalization of the TSP is the multiple TSP (mTSP), where a number m of salespeople have to serve a number of customers. The assignment-based formulation of the problems is:

$$min \sum_{i=1}^{n} \sum_{j=1}^{n} c_{ij} x_{ij}$$

Subject to:

$$\sum_{j=2}^{n} x_{1j} = m \tag{1}$$

$$\sum_{i=2}^{n} x_{i1} = m \tag{2}$$

$$\sum_{i=1}^{n} x_{ij} \quad j \in V \backslash \{1\} \tag{3}$$

$$\sum_{j=1}^{n} x_{ij} \quad i \in V \backslash \{1\} \tag{4}$$

$$\sum_{i \in S} \sum_{j \in S} x_{ij} \leq |S| - 1 \qquad \forall S \subseteq V \backslash \{1\} \tag{5}$$

$$x_{ij} \in \{0.1\}. \ \forall (i.j) \in A \tag{6}$$

where constraints (1) and (2) ensure that exactly m salespeople leave and return to the originating node (depot, which by default corresponds to node 1) and constraint (5) is the subtour elimination constraint.

The mTSP has been implemented for a wide range of applications; milestone examples include the mission planning problem in military operations, where a number of agents (e.g., unmanned aerial vehicles) located at a base must visit a number of goals,[50,51] the scheduling problem for school buses[123] with the objective to minimize the number of routes and the total traveled distance under the constraints of bus carrying capacity and allowed maximum time, and the crew scheduling problem.[124]

As mentioned, the VRP is a generalization of the TSP and regards a number of vehicles with the additional constraint that the total demands of all nodes in each particular route do not exceed the capacity of the vehicle. The VRP was initially introduced in 1959 as "the truck dispatching problem"[125] to describe a real-world application of routing a fleet of delivery trucks carrying gasoline from a bulk terminal to a number of dispersed gas stations. In 1964 an improved "greedy" heuristic algorithm was proposed for effective solution of the problem.[61]

6.4.1 Area Coverage Planning

The field area coverage problem described here concerns the reduction of the nonworking distance traveled by an agricultural machine following a system of parallel field-work tracks. In a specific system, for a given operating width the field-work tracks that the machine has to operate are well defined. What must be minimized is the distance traveled while moving the machine between any pair of consecutive tracks. By taking this into account, the length of each field-work track does not affect the minimization of the coverage problem since this length has to be covered in any case. What matters is the length of each turn executed by the machine, and the total length of these turns is a function of the order in which the machine works the field-work tracks.

Based on this, the notion of the field-work tracks is purely "topological", and this leads to the representation of a field-work track by a node in a graph and the representation of the field area coverage problem by the problem of graph traversal. As noted earlier, a graph[126] is defined by the set $G = (V.A)$ where $A = \{(i.j): i.j \in V; i \neq j\}$ and a cost matrix $C = (c_{ij}). (i.j) \in A$. In the simplified approach it is considered that each field-work track is represented by one node; consequently $V \equiv T = \{1.....|T|\}$.

6.4.2 Route Planning for Primary Units

The fieldwork pattern simulated in the case of traditional planning is the straight alternation pattern. In the case of a single-field operation, the field area is divided into a number of parts equal to the number of units. The applied rule states that the total effective distance (and not the number of the tracks) should be (about) the equal.

Next, a comparison between traditional and optimized planning of field operations carried out by one, two, three, and four identical units is presented. All relative time factors (i.e., blockages, preparing time, and operator's mistakes) are excluded from the operational time. The comparison is based on the traveled distance, which is divided into effective and nonworking (turnings and out-of-field travel) distance. The time elements (effective, noneffective) result from the corresponding traveled distances using average speeds. The averages used for the simulations were operating speed 2 m s^{-1} (7.2 km h^{-1}) as a mean typical driving speed for most field operations,[71] out-of-field traveling 1.8 m s^{-1}, maneuvering in smooth headland turnings 1.5 m s^{-1}, and maneuvering in steep headland turning 1.2 m s^{-1}.

In a field operation, headlands are created by the sequential passes that the agricultural machine has to perform peripheral to the field before or after (depending on the operation) the operation in the main field. Thus the headland width results from multiplying the effective operating width of the machine by the number of peripheral passes. For the simulated operations three headland passes were considered. The operation regarding headland area generation was common for both traditional and optimal planning.

- One unit: sequential passing on three headland tracks starting from the entry point into the field.
- Two units: the first two passes are carried out simultaneously by the units moving in parallel and the third pass is allocated to the unit with the shorter operational time. The operational time for the allocated route of each machine is not equal for all units, because there is a lower limit on the distance "entities" (the track lengths) which cannot be divided.
- Three units: all passes are carried out simultaneously by the three units moving in parallel.
- Four units: the passes are allocated to the (three) units with the shortest operational times.

To cover the range from large units to medium and small-size units, four machinery cases in terms of operating width and maneuverability (minimum turning radius) were considered (cases A, B, C, and D) (Table 36).

As test sites, four fields located at Research Centre Foulum (Denmark: N 56° 29′ 21.55, E 009° 34′ 59.40) were used (Fig. 59). The geometrical representation of the fields was based on shape files, including all the

Table 36 Four Machinery Cases Used in the Simulated Scenarios

	Size	Operating Width (m)	Minimum Turning Radius (m)
Case A	Small unit	1.5	3.5
Case B	Small/medium unit	3.0	3.5
Case C	Large/medium unit	3.0	5.0
Case B	Large unit	6.0	5.0

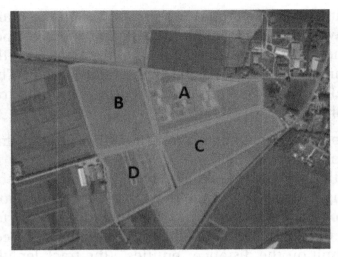

FIGURE 59 The farm and the four fields.

necessary information pertaining to the field as a geographic feature, provided by the GIS database of the Danish Ministry of Food, Agriculture, and Fisheries.

The areas of the fields are Field A 6.01 ha, Field B 5.65 ha, Field C 5.70 ha, and Field D 3.76 ha. Figs. 60–65 present the total operational time for each field separately as well for the whole farm (four fields) for the four machinery cases.

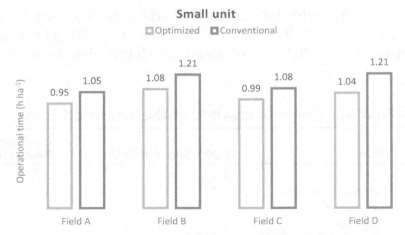

FIGURE 60 Operational time for each of the four selected fields for conventional and optimized execution of the work for a small unit.

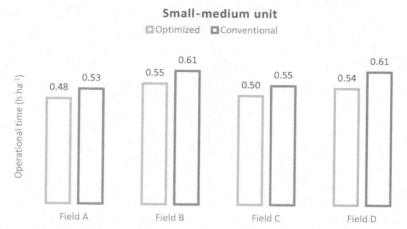

FIGURE 61 Operational time for each of the four selected fields for conventional and optimized execution of the work for a small/medium unit.

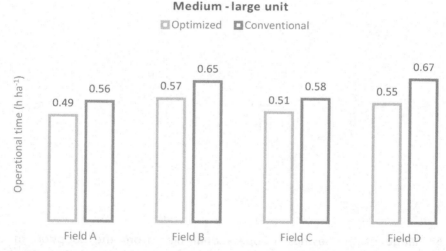

FIGURE 62 Operational time for each of the four selected fields for conventional and optimized execution of the work for a medium/large unit.

6.5 Performance Evaluation

In any dynamic operational system, three types of events can potentially deviate execution of the operation from the plan.

Urgent events. These events can result from an emergent operational situation, as for example a machine malfunction.

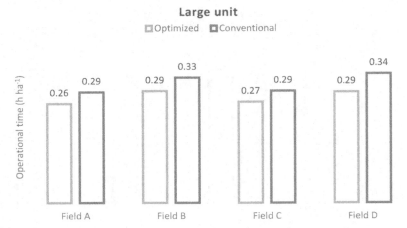

FIGURE 63 Operational time for each of the four selected fields for conventional and optimized execution of the work for a large unit.

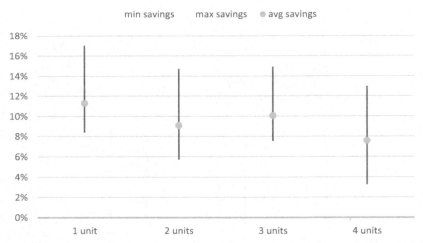

FIGURE 64 Range of the savings in operational time from the adoption of optimal fieldwork patterns in single-field operations.

Ordinary events. Although these events are expected, the exact time and place where they will take place are uncertain.

Discrete events. These events concern the starting and termination of a continuous process.

Fig. 66 shows a generalization of an event management system. More specifically, an event management system includes the following components.

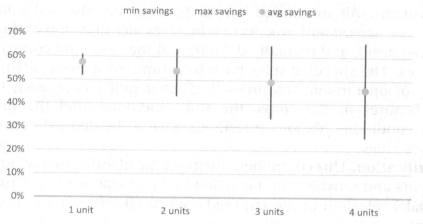

FIGURE 65 Range of savings in nonworking traveled distance from the adoption of optimal fieldwork patterns in single-field operations.

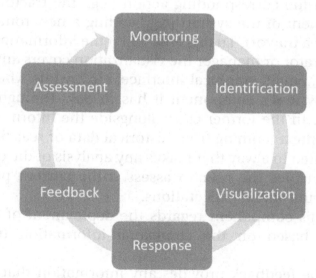

FIGURE 66 The general information flow in an event management system.

Monitoring. This includes the various data acquisition processes. As an example, in a field operation it is necessary to acquire data about machinery, i.e., operational state, rate of work, etc., any information on the soil conditions, e.g., moisture level, and information on the processed crops, i.e., maturity state, etc.

Assessment. All monitored events have to be assessed and documented. The assessment process first involves identification of the type of monitored event, and secondly definition of the rules and corresponding key factors. The specified rules have to comply with the availability and capacity of operational resources (i.e., labor and equipment), the biological features of the crops, the soil conditions, and the prevailing weather conditions. Various thresholds have to be defined according to quantified rules.

Identification. This component includes the identification of any event that occurs and provision of the action that corresponds to the rules, and the parallel selection of the required contextual information to enhance the event.

Visualization. After the identification process, the generated information must be given to the manager of the system, an appropriate person who has to take the corresponding action (e.g., the tractor operator), or another component of the system (e.g., adding a new route to the navigation system of a tractor). To this end, when the information is presented to a human operator or manager the visualization covers any visual report (e.g., messaging, email, graphical interface, etc.), while when it has to be delivered to a system's component it has to pass through a communication interface. In the former case, alongside the information itself the related event context (coming from historical data or real-time resources) has to be presented in a way that makes any analysis of the occurred event feasible. This enables the user to assess if the current performance is above or below the stated expectations.

Response. This component regards the deployment of a usually predefined action based on the contextual information that has been visualized.

Feedback. The feedback provides any information that can lead to a validation, a refinement, or a reconsideration of the planned processes.

7

Agriproducts Supply Chain Operations

7.1 Logistics Operations in Agricultural Production

Logistics is defined as a discipline that plans, implements, and controls the efficient and effective flow and storage of good, services, and related information from point of origin to point of consumption to meet customers' requirements.[127] This concept of management and logistics moves from the farm level to the supply-chain level, where the focus is also on customer requirements (Fig. 68). The customer could be the elevator that collects grain from the farms, the distribution centre of goods, or the biorefinery or biogas plant.

Logistics plays an important role in agriculture today, with challenges in terms of, for example, genetically modified crops, globalization, traceability, and local produce distribution. Some logistics techniques can be borrowed from the industry domain; but there is a need for adaptation (handling and storage of perishable produce, seasonal production and demand, timeliness constraints, food safety constraints in transportation)

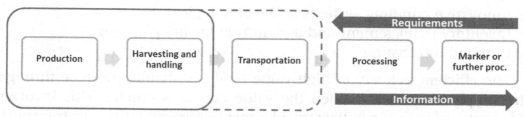

FIGURE 68 Concept of logistics chain.

Operations Management in Agriculture. https://doi.org/10.1016/B978-0-12-809786-1.00007-2

and some methodologies should be implemented ex novo. Agricultural logistics involve a number of objectives.

- To meet recent demands on machinery management in complex agricultural operations related to harvest, distribution, and transport of produce and byproducts (grain, biomass, slurry).
- To deploy state-of-the art technology for optimal management of on-farm, off-farm, and regional logistics operations.
- To develop methods and tools to improve the efficiency of logistics operations.
- To set up standard parameters for comparison of logistics operations.
- To study the use of intermodal logistics (e.g., switch between transport systems) to increase the performance of the operation.
- To optimize through a system approach the performance of the entire chain.

Agriculture involves operations like on-farm, off-farm, and regional logistics, service operation logistics and fleet management, biomass, forage, and grain supply chains, slurry and byproduct management, storage, and operation design. These agrofood and biomass logistics can profit from tools already used for industrial purposes.

- Discrete event simulation modeling.
- Linear, mixed, and integer programming.
- Analytical models and statistical tools.
- Vehicle route planning and logistics networks.
- Management resource planning and just-in-time (JIT) methodologies.
- Lean thinking applied to streaming of information and goods.
- Heuristic and scheduling tools.
- Innovative handling systems and technologies.
- Inventory management and agricultural facility planning.
- Demand forecasting.

The efficient delivery of high-value (and perishable) produce through the supply chain (also called the value chain) is crucial. This involves different performance indicators, like cost of transport, time for transport, and shelf life of the product for the final customer. Products range from local delivery of fresh highly perishable produce to global supply systems (e.g., global sourcing of biomass, grain, and meat).

One potential application relates to improving the information streaming along the supply chain. Efficient information flow can lead to improved supply-chain performance (JIT delivery of produce, better allocation of resources, machinery, and manpower, delivery cost reduction, etc.) without higher investments in transport or storage facilities. The traceability process is a typical field that profits from increased information sharing along the supply chain.

7.2 Agrifood Supply Chain Management

In general, in any supply-chain system there are five flow types, both upstream and downstream, in the value chain.

- Input and output (products and byproducts) material flows.
- Information flows.
- Financial flows.
- Processes.
- Energy flows.

All these flows are integrated in a system that includes numerous stakeholders: farmers, farmers' cooperatives, intermediaries, industrial partners, transporters, manufacturers, wholesalers, retailers, and consumers (Fig. 69).

Agrifood supply chains are characterized by unique features compared to normal supply chains, and thus customized managerial processes are needed.[128,129]

- High differentiation in products, both between various products and within a product itself.
- Seasonality in production operations and processes.
- Seasonality in production.
- Short life cycle in the case of perishable goods.
- Specific requirements in transportation.
- Specific conditions in storage.

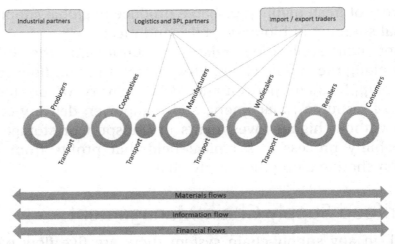

FIGURE 69 Stakeholders, interactions, and flows within typical agrifood supply chains.[130,131]

- Specific quality and safety requirements (traceability and visibility features).
- Need to comply with national and international regulations and legislations.

Fig. 70 presents the general hierarchy in a supply-chain management system in an agrifood supply chain[132] for the three decision levels, namely the strategic, tactical, and operational levels. At the strategic management level, the first family of decisions regards the product itself. The efficiency of a supply chain depends on the product design (this mainly concerns the postharvesting stage in the supply chain), the packaging, the manufacturing, and the potential value from the product disposal (e.g. recycling). For the environmental impact of a product, the design phase is the most significant for determining its ecoefficiency throughout its life cycle.[133]

7.3 Biomass Harvesting, Handling, and Transport Operations Management

Biomass logistics involves two parts of different scope. The first is the on-farm production of biomass as the initial step in the biomass supply chain. It is characterized as a low industrialized process, where planning

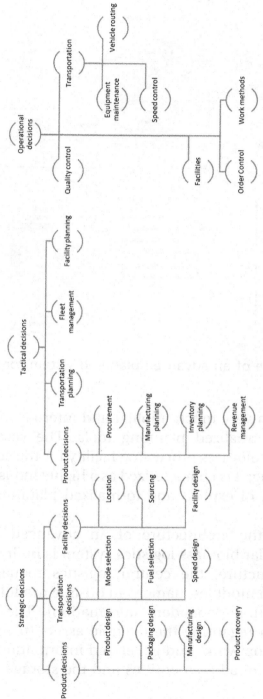

FIGURE 70 Hierarchy of the decision-making process in agrifood products.

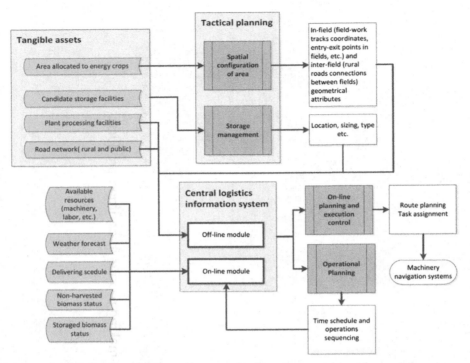

FIGURE 71 Architecture of an advanced planning system for a biomass supply chain.

and execution remain very much implicit and internal with only a sparse tradition of using formalized planning tools. The second part is the transport system and biomass processing facility for the specific bioenergy production/processing; it is characterized as a highly industrialized process with a long tradition of explicit and formalized planning and execution tools.

Fig. 71 presents the architecture of an advanced biomass supply network. The particular biomass logistics system is built around a central information infrastructure, the central logistics management system, which consists of two modules, namely an off-line module and an on-line module (specifying time-dependent information). The off-line module provides information regarding the tangible assets (plant processing facilities and the road network) and generated information concerning the spatial configuration of all area entities and the storage management by using tactical planning measures. More analytically, the system's entities include the following.

Tangible assets. The tangible assets forming the basic outline of the biomass network.

- Plant processing facilities.
- Area allocated to energy crop production.
- Candidate storage facilities (if any).
- Road network (rural and public).

Tactical planning. The first part of tactical planning regards the optimal spatial configuration of the fields making up the area allocated to the energy crops and involves a GIS-based method. The generated information in terms of in-field (field-work track coordinates, entry/exit points in fields, etc.) and interfield (rural road connections between fields) geometric attributes is transferred to the off-line module of the central logistics information management system. The second part concerns the decisions on identifying candidate storage facilities involving location, size, type, etc. The generated information is also transferred to the off-line module of the logistics information management system.

Biomass harvesting and handling operational planning. Operational planning must implement industrial engineering approaches (e.g., flow-shop scheduling, etc.) for time scheduling of operations and the interfield biomass harvesting/handling sequential operations planning. The operational planning uses as input the basic information regarding the tangible assets provided by the off-line module involving the central logistics information management system, and time-dependent information (machinery availability, delivering schedule request, weather forecast, crop condition, etc.) provided by the on-line module of the central information management system.

On-line planning and execution control. The next level of the system regards on-line planning and execution control of the generated operational and scheduling plans for the biomass supply chains which can be of various types (Fig. 72). This specifically regards the in-field operations (such as cutting and conditioning), the interfield

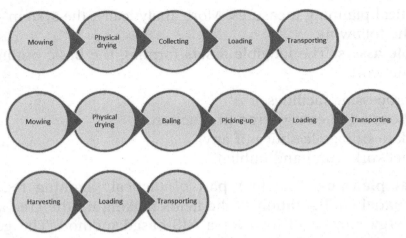

FIGURE 72 Green biomass supply chains.

machinery routing (navigation of machines between fields), and the transportation routing schemes for biomass. The plans have to be formatted and transferred from the central logistics information management system to navigation aiding systems.

8

Energy Inputs and Outputs in Agricultural Operations

8.1 Energy Usage in Agriculture

The input of machinery, fuels, pesticides, and fertilizers causes farming to depend heavily on energy,[134-137] so there is a continued effort to find ways to reduce inputs by applying new technologies and methods capable of reducing, for example, direct fossil energy consumption. As an example, Sørensen and Nielsen showed that new machinery systems and the use of reduced tillage methods could reduce the direct energy input by 18%–83% under different conditions.[87] However, it is important to consider not only the direct energy consumption but also the interrelated energy efficiency involving both direct and indirect energy. Indirect energy is associated with the production and distribution of inputs like fertilizers, seeds, machinery, and pesticides. Securing agricultural production methods with maximum net energy productivity and minimized impact is a key issue in future farming, especially in intensive cropping systems. Moerschmer and Gerowitt estimated that about 65% of total fossil energy input in, for example, wheat production was indirect energy, and similar results were derived from Danish cropping systems.[138,139] Hence estimations including direct as well as indirect energy use are required when elaborating on energy use and energy efficiency in alternative agricultural production systems, to avoid suboptimization and incorrect recommendations on technology use in farming operations. In this regard, it is interesting to consider whether less intensive tillage systems are more energy efficient than conventional systems. If one purpose of reducing fossil energy use is to reduce greenhouse gas (GHG) emissions, then it is necessary to include in the assessment the total emission of GHG in the compared systems. There may be other changes in GHG emissions related to changed soil preparation practices, and these

Operations Management in Agriculture. https://doi.org/10.1016/B978-0-12-809786-1.00008-4

187

may either reinforce or counterbalance the change in carbon dioxide (CO_2) emission from the energy use itself. Moreover, the changes in GHG emissions should be assessed in comparison with other interdependent environmental impacts from the cropping systems. Such complex relations throughout the product chain, across crop rotation, and between different environmental impacts call for a dedicated modeling approach and use of life-cycle assessment (LCA) methodology.

Other environmental impacts include the influence of the soil organic matter content. Intensive tillage of soils will decrease soil organic carbon content and increase the loss of CO_2 from soils.[140,141] As an example, it is reported that intensive tillage in the US Corn Belt has caused a reduction of the soil carbon content by 30%–50%,[142] leading to a significant emission of GHGs. In Denmark long-term experiments including removal of straw from fields show a net loss of carbon for each growing season.[143] Introducing less intensive tillage methods is expected to reduce the agricultural impact on the global CO_2 balance, as indicated by selected studies.[144]

8.2 Direct and Indirect Inputs

The direct energy used in field operations concerns the consumption of diesel by the tractors and other self-propelled machines involved. Consumption depends primarily on the engine size and engine load. The indirect energy of a product refers to its "embodied" or "embedded" energy, and includes the energy required to extract and process raw materials used for its manufacturing and the energy used during all the manufacturing activities required to make the finished product and transport it to the consumer.

8.2.1 Direct Energy Estimations

The estimation of direct energy for agricultural operations is based on a sequence of part-time operations including both effective working time and ancillary times such as turning and provision time. The average fuel requirement per hour is estimated as a function of engine size and average engine loads during the operations in question. For example, the engine load can be set to 60%, representing the same fraction of the maximum power of the tractor and based on the measurements from farm analyses.[145] For combine harvesters the average engine load is higher and can be set to 80%. The average fuel consumption is estimated as:

$$l(f)_j = \frac{x \cdot P_m \cdot f_e}{1000 \cdot d}$$

where $l(f)_j$ is the average fuel consumption in $l\,h^{-1}$ for the j'th field operation, x is the engine load, P_m is the maximum power of the tractor in kW, f_e is the assumed fuel efficiency in g kWh^{-1} (set to 250 g kWh^{-1}), and d is the density in kg l^{-1} (set to 0.84 kg l^{-1}). The estimation of direct energy use for specific operations scenarios is calculated as:

$$DE = \sum_j \frac{1}{CAP} \cdot (l(f)_j \cdot EC(f) + l(o)_j \cdot EC(o))$$

where DE is the total direct energy allocated to a sequence of specific operations (in MJ), CAP_j is the capacity of the field operation j (in ha h^{-1}), $l(f)_j$ is the average fuel consumption (in $l\,h^{-1}$), $EC(f)$ is the norm energy coefficient set at 47.8 MJ kg^{-1}.[8] $l(o)_j$ is the lubrication oil of motors assessed at 2% of the diesel consumption,[146] and $EC(o)$ is the norm energy coefficient set at 43.8 MJ kg^{-1}.[8]

8.2.2 Indirect Energy Estimations

Embodied energy, or "embedded energy," includes the energy required to extract raw materials from nature plus the energy used in primary and secondary manufacturing activities to provide a finished product and transport it to the consumer.[147] Estimates of the embodied energy of the different machinery items used in agriculture are based on the total mass of each item multiplied by energy coefficients, which represent the sum of raw material extraction, manufacturing, transport from the factory to the farmer, installation, and maintenance and repair. The annual embodied energy input to in-field operations is allocated by amortising this over the expected working life of the machinery item. Embodied energy or indirect inputs are difficult to quantify, as these energy components are varied and complex and there is no standardized methodology for assessing them. Several authors suggest different valuations.[147–149] Setting system borders is even more difficult for indirect energy input than it is for direct energy input.[150] Doering showed that a number of different methodological approaches are used in the determination of indirect energy.[151] In terms of indirect energy for tractors, Bowers gave an estimate of 95.6 MJ kg^{-1}, while Nagy proposed 61.2 MJ kg^{-1}.[73,152] Also, technology changes over time and affects the efficiency of machines. Dyer and Desjardins proposed a reduction of earlier indirect energy estimates by 23% to account for technology improvements and use of diversified materials in tractor construction.[153] Correctly, the estimation of indirect energy demand for tractors and other self-propelled machines must be based on a precise breakdown of the material composition of modern tractors but using established energy coefficients for individual materials and later reusage of scrap.

The machinery items for different tillage systems fall into two main categories: tractors, and tractor-powered implements like ploughs, seed drills, etc. (Table 37).

Using the derived figures for embodied energy per unit of mass for the different machinery categories in Table 37, the contribution of machinery items to the total energy demand can be estimated depending on the level of mechanization, crop area, etc.:

$$IE = \sum_i \frac{W_i \cdot EC_i}{UL_i} \cdot UT_i$$

where IE is the total embodied energy allocated to a specific operation (MJ), W_i is the mass of the machine item i (kg), EC_i is the assessed indirect energy coefficient per unit mass (MJ kg^{-1}), UL_i is the assumed lifetime use (h), UT_i is the machinery time allocated to the operation in question (h), i is an index for machinery items, UL_i is based on assumptions.[162]

8.2.2.1 Machine Weights

The annual allocated weights of the applied field machines used for the indirect energy calculations are estimated as evenly distributed over the economic life of the individual machine and also over the different crops in the cropping system. The economic life depends on the annual utilization of the machine, as high utilization will reduce the economic life span. In addition, a maximum economic life is estimated for each

Table 37 Material Requirements and Energy for Self-Propelled Machines and Implement Manufacturing

Materials[a]	Self-Propelled Machine Manufacturing			Implement Manufacturing		
	g kg^{-1}[b]	MJ kg^{-1}[c]	MJ	kg kg^{-1}	MJ kg^{-1}	MJ
Aluminium, alloy 85% recycling	5.20	191	0.9932			
Copper, primary 85% recycling	0.92	100	0.092			
Fuel and oil[d]	49.50	46.8	2.3166			33.00
Glass, laminated 85% recycling	2.30	26.2	0.0603			
Cast iron, unalloyed 85% recycling	590.00	27.6	16.284			
Plastic[e]	11.00	90	0.99			
Steel, low-alloyed 85% recycling[f]	0.23000	33	7.59			
Tire rubber, synthetic 100% recycling	0.11750	110	12.925	1	33	
Total materials production			41.25			33.00
Process energy[g]			24.40			14.40
Total manufacturing			65.65			47.40
Transportation[h]			1.29			1.29
Repairs and maintenance[i]			17.07			14.22
Total			86.01			62.91

[a]Material characteristics based on various authors.[154–157] The amount of recycling for the different components is based on EU requirements on January 1, 2006.[156]
[b]Component distribution (g per kg in total) based on average values from Heller et al. and Cho and Han.[158,159]
[c]Energy coefficients for raw material production are derived from Bowers and Audsley et al.[73,160]
[d]Fuel and oil consists of 84% diesel fuel and 16% engine oil.[158]
[e]Plastic breakdown: 25% polyurethane, 37.5% polypropylene, 37.5% acrylonitrile butadiene styrene.[158]
[f]Typical composition as described by Nemecek and Erzinger.[161]
[g]The energy for manufacturing and assembly based on Heller et al.[158]
[h]Machinery items are assumed to be transported 1000 km by rail and 200 km by truck from the factory to the farmer, requiring 0.001,187 MJ kg per km (rail) plus 0.0,005,377 MJ kg km^{-1} (truck), equaling 1.29 MJ kg^{-1}.[160]
[i]Repair and maintenance was set to 26% of the total manufacturing energy for tractors above 50 kW.[160]

machine as a limit not to be exceeded even if the annual utilization of the machine is low.[162]

The initial machine weights are based on available machine specifications.[163] To estimate machine weights over a range of machine sizes, regression models for weights as a function of sizes are derived and averaged over makes available on the market. As an example, the functional mass relationship for a combine harvester is as follows:

$$a = 4647 + (1.29 \cdot \beta \cdot 1000)$$

where a is the estimated machine mass in kg and β is the estimated machine size in terms of working width in meters. The mass estimators for other machine types are derived in a similar way.

8.2.2.2 Operation and Task Times
As mentioned, it is necessary to estimate UT_i as the machinery time allocated to a specific operation to estimate energy requirements for that operation. The labor and machine input for targeted operations uses a modeling approach from general work management.[87,164,165] Following this approach, labor and machine input are estimated as a function of parameters, such as field dimensions, working speed, working width, and transport distance:

$$A = \left(\frac{600 \cdot h}{v \cdot e} + \frac{p \cdot b \cdot n}{e \cdot (1+a)} + k + s \cdot h\right) \cdot (1+q) \cdot h^{-1}$$

where A is the labor requirement in minutes, h is the size of the field in ha, v is the working speed in km h^{-1}, e is the effective working width in m, p is the time for turning in minutes per turning, b is the field width in m, n is the number of turnings per pass (normally $n = 2$), a is a model parameter dependent on field shape and travel pattern ($a = 1$ in the case of driving back and forth in the swath), k is the turnings on headland in minutes per field, s represents the stochastic interruptions due to crop and soil attributes, adjustments, control, and tending of machine in min ha^{-1}, and q is an assessed rest allowance time amounting to 5% additional time in minutes.

Ancillary work elements included filling of the hopper with seeds and fertilizer, and were modeled as follows:

$$I = \left(\frac{c \cdot z}{100} + \frac{m \cdot z}{w}\right) \cdot (1+q)$$

where I is the labor requirement in min ha^{-1}, c is the filling capacity in minutes $(100 \text{ kg})^{-1}$, z is the dosage in kg ha^{-1}, m is preparation and termination of the filling operation in minutes per filling, w is the capacity of the hopper in kg per filling, and q is an assessed rest allowance time amounting to 5% additional time in minutes.

The remaining prerequisites included the selection of a 10 ha field and a field shape configured with a field length of double the width (224 × 447 m). The working speed was selected based on the average measured speed from farm analyses.[87] Machine capacity was modeled as the in-field capacity incorporating time for auxiliary turnings, loading of seeds, adjustments, compulsory stops, and personal allowances.

8.3 Assessment Tools

8.3.1 Life-Cycle Assessment

The globally accepted framework for performing an LCA is described by the ISO LCA Standard. The first version of the ISO LCA Standard was published in 1997. The LCA procedures are part of the ISO 14,000 environmental management standards: ISO 14,040:2006 is written for a managerial audience and provides the "principles and framework" of the standard, while ISO 14,044:2006 provides the "requirements and guidelines" for a practitioner.

A summary of the LCA at a high level is given in Fig. 73, where the four basic phases of the assessment process are presented, namely goal and scope definition, inventory analysis, impact assessment, and interpretation.

The **goal and scope** are statements of intent for the study at hand and provide the study design parameters. In the "goal" phase the system's items are defined:

- the product system at hand
- the system boundary

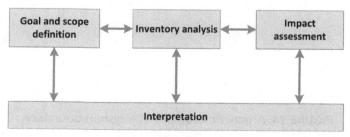

FIGURE 73 Life-cycle assessment framework.

- the functional unit
- the definition of the inventory input and output parameters
- the inventory method or methods to be implemented.

A product system, based on ISO 14,040:2006, is a collection of processes that provides a certain function. A product system comprises a number of constituent subsystems of processes and flows. The system boundary distinguishes which of these subsystems constitute part of the study at hand and which do not, based on the stated goal and scope of the study. Fig. 74 presents a generic example of a product system and a chosen system boundary.

The functional unit is a quantitatively defined measure of this function, making it possible to relate it to the relevant input and output parameters of the product system. In other words, the functional unit bridges the function of the product system with the inputs (items such as energy and resource use) and outputs (items such as emissions or waste produced). It is worth noting that for comparisons of alternative product systems with

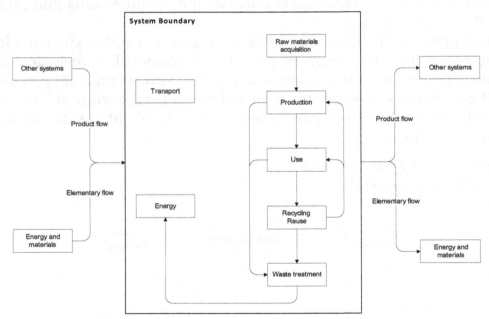

FIGURE 74 A generic example of a system boundary.

the same function, the selected functional unit must be consistent across all alternatives. A comparative study can only be conducted if the study design allows each product to be compared across the same goal and scope, functional unit, and system boundary.

The **inventory analysis** provides the data needed to perform an analysis that meets the stated goal and scope. Fig. 75 presents the steps required for the inventory analysis process. This process might be

FIGURE 75 Inventory analysis workflow.

Table 38   Impact Categories Considered in LCA

General Distinction	Impact Category
Input related	Depletion of abiotic resources. Consumption of nonrenewable resources such as water and crude oil.
	Land use. The use (occupation, conversion, transformation) of areas for human-related activities such as agricultural production, road networks, housing, mining, etc.
Output related	Climate change, or global warming potential, expressed in CO_2 equivalents derived from the rate of Carbon dioxide (CO_2), Methane (CH_4), Nitrous oxide (N_2O), chlorofluorocarbons, and other GHGs released into the environment. These gases absorb the infrared rays that reflect off land and water, preventing their escape from the Earth's atmosphere and consequently causing the heating up of the atmosphere.
	Stratospheric ozone depletion.
	Human toxicity/ecotoxicity.
	Photooxidant formation.
	Acidification. Reduction of soil and water pH, which affects animal and plant life.
	Nutrification (eutrophication). Release of nutrients, mainly nitrogen and phosphorus, from sewage outlets and fertilized farmland causes nutrient enrichment. This results in changes in species composition in nutrient-poor habitats and algal blooms causing conditions of lack of oxygen and fish death.

repetitive, since as depicted at Fig. 73 the whole process within the LCA framework could include reconsideration of the various phases, and all these phases are also reconsidered according to the then-current defined goal and scope of the study.

In the **impact assessment** phase the inventory results are transitioned to climate change effects. In this stage the impact categories to be studied, the methodology of the impact assessment to be followed, and the subsequent interpretation to be used are selected. The impact indicator for each category (always per functional unit) is given by:

$$I_i = \sum_j E_j \cdot CF_{ij} + \sum_k R_k \cdot CF_{ik}$$

where I_i is the impact indicator for category i, E_j the release of emission j, R_k the consumption of resource kk, and $CF_{i(j \text{ or } k)}$ is the characterization factor for emission j or consumption of resource k, respectively, contributing to impact category i, which represents the potential of a single emission or resource consumption to contribute to the respective impact category.[166]

The values of the impact category indicators for a system under study is called the "environmental profile" of the system. Table 38 lists the impact categories considered in an LCA study.[167]

Finally, in the **interpretation** phase the results of the study are put into perspective and any recommendations for improvements of system performance in terms of reduced environmental impact are stated.

9

Advances and Future Trends in Agricultural Machinery and Management

9.1 Robotics

Labor contributes between 20% and 40% of the operational cost in agricultural production.[168,169] Furthermore, especially in countries where wages are low, farm employment is not on any kind of permanent basis. This as a consequence results in low levels of expertise, leading to lower productivity and lower performance quality. Automation and control technologies and information and communication technologies (ICTs) such as wireless networks, global positioning systems (GPS), geographic information systems, farm management information systems (FMIS), etc. have brought significant improvements in productivity, offering management advantages through applications such as sensing, data mining, and analytics. As a next step in this development, autonomous machines and robotic technologies can further advance agricultural systems' productivity. In these changes in the model of agricultural production, the main factor being increased is the capacity of machines as a result of their physical optimization (advancement in the mechanical technologies) and increased efficiency due to the embodied automation and information technologies. As parallel consequences, labor cost is also decreased and environmental impact has been mitigated, due to controlled inputs (e.g., agrochemicals). The latter improvement also arose from the wider implementation of the precision agriculture (PA) principles.[170]

Robotics in agriculture is not a new phenomenon; in controlled environments it has a history of over 20 years. However, with the latest increase in computational power combined with a cost reduction, robotics applications are spreading. In recent years the development of autonomous machines in agriculture has seen increased interest, and many researchers are developing autonomous vehicles for agricultural operations. The prime concept is to use many small, efficient autonomous machines in place of large tractors.[171] These autonomous vehicles should be capable of working 24 h a day all year round, in most weather conditions, and have the embedded intelligence to behave sensibly in a natural or seminatural environment over long periods, unattended, while carrying out a useful task.

The main benefits of autonomous and intelligent agricultural robots are improving repeatable precision, efficiency, and reliability and minimizing soil compaction. The robots have potential for multitasking, sensory acuity, and operational consistency as well as suitability for different operating environments, interactions, physical formats, and functions (Fig. 76).

Fig. 76 shows the various categories of robots found in the general industrial and service sectors. Robots are categorized according to the operating environment (ground, air, water, space, inside the human body), the interaction and collaboration type (programmed, teleoperated, supervised, collaborative, autonomous), their physical format (robotic arms, exoskeletal, metamorphic, micro, humanoid), and their function (assembly, area process, transport, inspection, manipulation).

Researchers are now focusing on different farming operational parameters to design autonomous agricultural vehicles, as conventional farm machines are crop and topology dependent. To date, agricultural robots have been researched and developed principally for harvesting, chemical spraying, picking fruit, and crop monitoring. Robots like these are perfect substitutes for manpower to a great extent, as they deploy unmanned sensing and machinery systems.

A number of field operations can be executed by autonomous agricultural robots, giving benefits over conventional machines. These autonomous platforms are used for cultivation and seeding, weeding, scouting, application of fertilizers and chemicals, irrigation, and harvesting. For harvesting, researchers are developing rational and adaptable robotic platforms for picking cucumber,[172,173] tomato,[174,175] pepper,[176–178] strawberry,[179–181] eggplant,[182] melon, watermelon,[183,184]

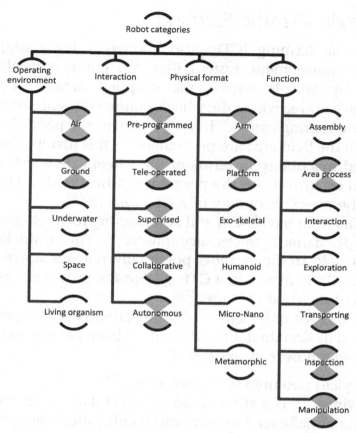

FIGURE 76 Categories of robots. In gray are the robotic types implemented in the agricultural production domain.

asparagus, cabbage, radish,[185,186] rice and paddy fields,[187,188] mushrooms,[189] cherry,[190] apple,[191,192] and citrus fruit.[193,194]

One of the important challenges of using robots for weed control is identifying weeds from crops. In plant protection and weed control researchers are developing autonomous robotic platforms for weed destruction.[195–199] Also, transplanting and seedling production involve operations such as seeding, thinning, grafting, planting cuttings, transplanting, and others. Some of these operations are automated or fully robotized.[200–204]

9.2 Controlled-Traffic Farming

Controlled-traffic farming (CTF; also known as the tramline farming system) is a management system that can completely eliminate soil compaction by wheels within the cropped area.[205] CTF restricts compaction to the area where the wheels contact the soil surface, and thus provide a loose rooting zone.[206] In a CTF system the parallel wheel tracks created within the field area are permanent[207]; this involves modifications to the applied machinery in terms of the wheel axle width to allow the wheels to run exclusively on the permanent wheel tracks. This is a factor that affects the overall economy of a CTF system.[208]

CTF benefits from navigation and autosteering technologies, since they follow the permanent tracks accurately.[209] The trafficked area in conventional field traffic systems potentially reaches up to 80% of the total field area. In contrast, in a CTF system the trafficked area is limited to 30%–60% of the total field area.[210]

CTF has a long list of benefits, including conserving the productivity of the system and its sustainability, which have been have been documented in various research over recent decades.[211,212]

- Increased yield potential of various crops.[208,213,214]
- Energy savings.[215] The elimination of wheel damage on the cropped area can potentially lead to substantial cultivation energy savings, ranging from 37% to 70%.[208]
- Fig. 77 presents the reduction ranges of various elements associated with sustainability that derive from implementation of CTF systems instead of farming systems based on conventional (noncontrolled) traffic. In the case of fertilizers, pesticides, seeds, and fuels, the reduction regards any indirect impact associated with each of these elements.

9.3 Precision Farming Management

The future of agricultural bioproduction systems calls for an unprecedented level and sophistication in doing the right thing in the right place at the right time. This will require a high degree of embedded intelligence in machinery, advanced operations management systems in constant

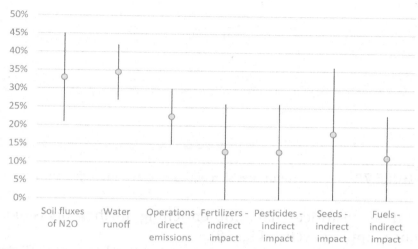

FIGURE 77 Reduction range in various sustainability-related elements derived from implementation of controlled-traffic farming systems compared to conventional traffic farming systems.

communication with mobile units, the implementation of automated systems, advanced up-to-date decision support systems, automatic data acquisition as an integral part of all machine operations for traceability/documentation, and production processes aligned with economic and environmental requirements. All this comes together in the concept of PA, which embodies the future development pathway for agriculture and other types of bioproduction.

Roughly speaking, PA is about doing the right thing in the right place, at the right time, and in the right way, indicating that its implementation is based on the application of technologies that define "right."[216] In more scientific terms, PA can be defined as an agricultural production management system that takes into account the spatial and temporal variability in fields and crops based on the use of ICTs.[217]

As a management system, PA is a closed-loop operation system working from collection of data to data interpretation, decision-making, performance of the designated actions, and finally evaluation of the system outcome, derived from application of the decisions and the reconsideration of the decisions taken. During each cropping cycle, the collected data is recorded and stored in databases (usually called libraries)

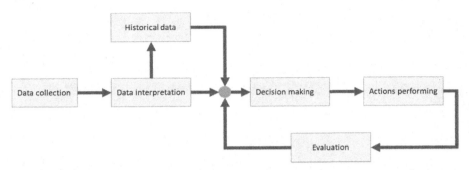

FIGURE 78 The concept of precision agriculture as a management system.

to be used as historical data for enriching further decisions in the subsequent cropping cycles (Fig. 78).

Precision farming is recognized as the enabler of efficient and sustainable agriculture in the future. Two directly noticeable side benefits of its implementation are a reduction of the environmental impact of production (due to the reduction in input material) and the generation of valuable information to be used in traceability of products. Precision farming embodies the development of new integrated concepts involving advanced planning and control systems and semiautomated or fully automated smart and robust machines applicable for biosystems.[4] These include production units complying with a range of sustainability indicators like minimized resource input, product quality, and environmental impact. As such, PA focuses on developing technologies and techniques which improve production efficiency and measuring, modeling, and limiting the environmental impacts of agricultural production systems. Specifically, this involves integration of site-specific application of fertilizers, pesticides, and water and operations management (farm management information systems, decision support systems, web-based approaches, etc.), since site-specific applications require automated information exchange between different components of the production system (e.g., current state of crop and soil) as a prerequisite for the algorithmic generation of decisions for optimum applications.

The concept of PA comes from the early years of organized agricultural production, when the implementation this principle was possible due to the small size of fields at that time. In these small farms, a farmer was able to walk all over the field area and observe any variations in terms of either crop emergence or crop growth, and take actions such as locally placing more seeds or more fertilizer. However, any decision leading to such actions was based on direct observation and limited to any "stored" knowledge gained from observations of previous cropping years. The inability to store and organize this knowledge continuously increased with the increase of the size of the field area during the mechanization phase in agricultural production. This phase, alongside with the underlying economies of scale, forced the homogeneous treatment of large field areas, although variations of yield and soil properties become more intense in larger field areas. Furthermore, the existing technology was not able to support any type of PA application.

The first technology applied in agricultural operations that provided the basis for the implementation of PA was the GPS. The next step was the introduction of yield-mapping technologies, initially in arable farming and in a stepwise manner in open-air horticulture and orchards.

Studies demonstrate that usefulness and ease of use are central aspects for technology adoption, provided that these aspects do not cause a significant increase in the production cost.[218] Other drivers of adoption include farm size, cost reduction or higher revenues to acquire a positive benefit/cost ratio, total income, land tenure, farmers' education, familiarity with computers, access to information (via extension services, service providers, technology sellers), and location. Also key to the adoption of PA are the quantified cost/benefits derived from the use of various technologies, such as areas of machine control, guidance, auto-section control, remote sensing, variable seed rate, and nutrient management.

Despite the fast development of PA technologies over recent decades PA and increasing numbers of verified use cases, adoption of these systems by farmers was considerably slower.[219,220] For a new technology to be adopted, one or more of the following criteria have to be met:

- it provides a proven economic benefit
- it provides crucial advantages compared to the existing technology that it replaces
- it is less complicated that the existing technology
- it is robust and reliable
- it is well supported by servicing and repairs.

One of the reasons for the low rate of adoption is the establishment cost of the related technologies. This led to a new generation of third parties providing site-specific technologies such as yield mapping and variable rate applications.[221]

Roughly, the cost of PA technologies derives from:

1. depreciation of the equipment
2. cost for training
3. variable cost for annual data analysis.

9.4 Satellite Navigation

The main satellite systems implemented in navigation tasks are listed below.

- USA: NAVSTAR GPS (Navigation System with Time and Ranging Global Positioning System). This system is composed of a network of 24 satellites placed into orbit and has been developed by the US Department of Defense and the US Department of Transportation.
- Russia: GLONASS (Global Navigation Satellite System/Globaluaya Navigatsionnaya Sputnikovaya Sistema). Another national military system, but both NAVSTAR GPS and GLONASS are internationally available for commercial and private use.
- Europe: GNSS (Global Navigation Satellite System). This is a civilian satellite navigation system.

Nowadays, there is an enormous growth in the applications of GPS in various domains. Selected applications include the following.

- In-vehicle navigation systems. These technologies improve the in-vehicle experience,[222] increase safety,[223] and reduce the burden on the driver by reducing his/her need to engage in wayfinding in unfamiliar environments. Based on a satellite navigation system, the location of the vehicle is identified in relation to the desired destination input by the driver. The systems offer both visual and auditory directions in almost real time.
- Fleet management. This uses GPS to reach and store location of origin, location of destinations, trip length and duration, various time stamps during traveling, travel modes, and activities. This system can

potentially completely replace the traditional methods of activity diaries and provide data that can be used for further analysis of transport systems.[224,225]

9.5 Farm Management Information Systems

When managing a farm today, the focus is very much on both economic efficiency and interaction with the surroundings in terms of environmental impact, timely delivery, and documentation of quality and growing conditions.[139] This is influenced by external entities (government, public), putting increased pressure on the agricultural sector to alter production to focus on quality and sustainability.[226] Key issues here involve provisions and restrictions in terms of, for example, fertilizers and agrochemicals and other incentives that encourage the farmer to adopt sustainable production. In general, the future requirements and challenges of arable farming systems (climate change, environmental impact, increased production, legislation/traceability, low use of technology, etc.) call for unprecedented sophistication in doing the right thing in the right place at the right time and operation optimization on a system level. Such a concept will apply to any farm operation, such as soil management, seeding, planting, fertilization/irrigation, harvesting, etc. It requires a high degree of embedded intelligence in machinery, advanced FMIS in constant communication with mobile units, the implementation of automated systems, advanced up-to-date decision support systems, automatic data acquisition as an integral part of all machine operations for traceability/documentation, production process aligned with economic and environmental requirements, etc.[227]

Over the last decades the productivity development in agriculture has incrementally moved from scaling of assets to optimization of assets (i.e., the focus is not the maximization of production but rather the maximization of profit). Agricultural machinery and overall farm enterprises have grown in size and value over the years, and a higher degree of input/output management has become essential for the farmer's profit. However, current farm and operation optimization efforts have more or less exclusively been directed toward management of single operations and neglected the overall systems approach.

There is an urgent need to evolve farm and operations management to allow current automated systems with crude management information support to take full advantage of the technological opportunities and integration. Recent advances in research and development in ICTs, robotics, processing, and operations management tools and the application of these advances have enabled a new paradigm shift in higher production efficiency and sustainability through the use of integrated processing, planning, and control systems. The machine systems or production units will receive control and guidance instructions from managers or automated decision-making systems (FMIS) using on-line sensing, meteorological data, historical information from the management system, and other information sources. Plan generation and execution will be linked in a system monitoring effects of actions, unexpected events, and any new information that can lead to to a validation, refinement, or reconsideration of the plan or goal.

Information is produced from many sources and many sites, putting high demands on suitable information systems. McCown stated that in an information system the focus should be on learning farmers' behavior.[228] Software developers must understand the farmers and interact closely with them in a user-centric approach.

The enhancement of FMIS is shown to be influenced more by prevailing business factors and drivers than specific farming activities.[229] Plans must be conditional, such that data from observations, databases, sensors, and tests can be used to update the a priori plans. Management information systems (MIS) are an integral part of an overall management system and support tools such as enterprise resource planning (ERP), information systems, etc. ERP is an industry concept involving a range of management activities supporting business processes within the production system. The management systems support planning processes on multiple levels and allow for identification of key performance indicators.[230] Typically, ERP is directly connected with information systems in the form of databases, and includes applications for the finance and human resources aspects of a business.

MIS is the basic concept of information systems with the objective of analyzing other systems dealing with the operational activities in the production system. By this notation, MIS is defined as part of the overall

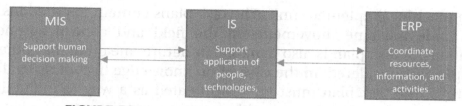

FIGURE 79 Concept of management information systems.

planning and control for the management of humans, technologies, and procedures inherent in the production system. As part of scientific management, MIS is tailored to the automation or support of human decision-making.[231] Fig. 79 shows the conceptual structure of the different management systems in a production system or enterprise.

By following the above described framework and notation, An FMIS is determined as a system for the collecting, processing, storing, and disseminating of data needed to execute the operations functions of a farm.

Mechanization has increased agricultural productivity considerably in recent decades. This development is current being integrated with advanced automation and extensive use of embedded ICT systems. These new technologies enable collection of detailed site-specific information during operation of field machinery (conventional, such as a tractor with an implement, or autonomous vehicles, such as robots). This contributes to reduced resource input and environmentally sound and quality-optimized production via targeted decision support systems or directly via on-line control. Planning and formulation of tasks in agriculture involve expected time schedules on the basis of the predicted crop development, weather forecasts, etc. The task specifications are transferred to the tractor/implement for control or manual/automatic adjustment of implements. When executed work differs from the formulated plans, corrective measures are initiated. The final work result is recorded and documented, and the data obtained is stored for learning and use in connection with new loops of planning or control.

Within the concept of the task management function described above, this function will enable the farmer to schedule and control the tasks relevant to given field operations. The key output from the planning processes is the formulated operation and task plan, which is transferred to the machine/implement unit. The task plans contain instructions that both guide machine movements in the field and control agronomic operations. A task plan is also dynamic in nature, meaning that time will be explicitly considered; in the case of no knowledge becoming available over time, the task plan must be reformulated as a way to maintain an optimal operation. Concurrent with machine and operations planning, the system must be able to analyze and predict the evolution of system states (field and crop development, machine performance, etc.) as input for the operations planning.

10

Appendices

Abbreviations

AC Annual cost
AFE Area-based field efficiency
CC Capital consumption
CPM Critical path method
CRF Capital recovery factor
DFE Distanced-based field efficiency
EFC Effective field capacity
ERP Enterprise resource planning
FC Fixed costs
FE Field efficiency
GNSS Global navigation satellite system
HYVs High-yielding varieties
IC Interest charge
ICC Instantaneous centre of curvature
ICT Information and communication technology
LC Lubrication consumption
LCA Life-cycle assessment
LP Linear programming
LPG Liquefied petroleum gas
PERT Program evaluation and review technique
PTO Power take-off
RMC SP Repair/maintenance costs self-propelled
RV Remaining value
SFC Specific fuel consumption
SWOT Strengths—weaknesses—opportunities—threats
TFC Theoretical field capacity
TSP Travelling salesman problem
VRP Vehicle routing problem

Operations Management in Agriculture. https://doi.org/10.1016/B978-0-12-809786-1.00010-2
Copyright © 2019 Elsevier Inc. All rights reserved.

Table I Draft Parameters and Expected Range in Drafts Estimated by Model Parameters for Tillage and Seeding Implements

Implement	Field Speed S (km h⁻¹)		Field Efficiency E_f (%)		Width Units	Machine Parameters			Soil Parameters		
	Range	Mean Value	Range	Mean Value		A	B	C	F_1	F_2	F_3
Major Tillage Tools											
Subsoiler/manure injector											
Narrow point	5.0–10.0	8.0	70–90	85	tools	226	0.0	1.8	1.0	0.70	0.45
30 cm winged point	5.0–10.0	8.0	70–90	85	tools	294	0.0	2.4	1.0	0.70	0.45
Moldboard plow	5.0–10.0	7.0	70–90	80	m	652	0.0	5.1	1.0	0.70	0.45
Chisel plow											
5 cm straight point	5.0–10.0	8.0	70–90	85	tools	91	5.4	0.0	1.0	0.85	0.65
7.5 cm shovel/35 cm sweep	5.0–10.0	8.0	70–90	85	tools	107	6.3	0.0	1.0	0.85	0.65
10 cm twisted shovel	5.0–10.0	7.0	70–90	80	tools	123	7.3	0.0	1.0	0.85	0.65
Disk harrow, tandem											
Primary tillage	6.5–11.0	10.0	70–90	80	m	309	16.0	0.0	1.0	0.88	0.78
Secondary tillage	6.5–11.0	10.0	70–90	80	m	216	11.2	0.0	1.0	0.88	0.78
Disk harrow, offset											
Primary tillage	6.5–11.0	10.0	70–90	80	m	364	18.8	0.0	1.0	0.88	0.78
Secondary tillage	6.5–11.0	10.0	70–90	80	m	254	13.2	0.0	1.0	0.88	0.78
Disk gang, single											
Primary tillage	6.5–11.0	10.0	70–90	80	m	124	6.4	0.0	1.0	0.88	0.78
Secondary tillage	6.5–11.0	10.0	70–90	80	m	86	4.5	0.0	1.0	0.88	0.78
Field cultivator											
Primary tillage	6.5–10.5	8.0	70–90	85	tools	46	2.8	0.0	1.0	0.85	0.65
Secondary tillage	6.5–10.5	8.0	70–90	85	tools	32	1.9	0.0	1.0	0.85	0.65

Row crop cultivator												
S-tine	5.0–11.0	8.0	70–90	80	rows	140	7.0	0.0	1.0	1.0	0.85	0.65
C-shank	5.0–11.0	8.0	70–90	80	rows	260	13.0	0.0	1.0	1.0	0.85	0.65
No-till	5.0–11.0	8.0	70–90	80	rows	435	21.8	0.0	1.0	1.0	0.85	0.65
Minor Tillage Tools												
Rotary hoe	13.0–22.0	19.0	70–85	80	m	600	0.0	0.0	1.0	1.0	1.0	1.0
Spring-tooth harrow	8.0–13.0	11.0	70–90	85	m	2000	0.0	0.0	1.0	1.0	1.0	1.0
Roller packer	8.0–13.0	11.0	70–90	85	m	600	0.0	0.0	1.0	1.0	1.0	1.0
Roller harrow	7.0–12.0	10.0	70–90	85	m	2600	0.0	0.0	1.0	1.0	1.0	1.0
Land plane	7.0–12.0	10.0	70–90	85	m	8000	0.0	0.0	1.0	1.0	1.0	1.0
Seeding Implements												
Row crop planter, prepared seedbed												
Mounted												
Seeding only	6.5–11.0	9.0	50–75	65	rows	500	0.0	0.0	1.0	1.0	1.0	1.0
Drawn												
Seeding only	6.5–11.0	9.0	50–75	65	rows	900	0.0	0.0	1.0	1.0	1.0	1.0
Seed, fertilizer, herbicides	6.5–11.0	9.0	50–75	65	rows	1550	0.0	0.0	1.0	1.0	1.0	1.0
Row crop planter, no-till												
Seed, fertilizer, herbicides												
1 fluted coulter/row	6.5–11.0	9.0	50–75	65	rows	1820	0.0	0.0	1.0	1.0	0.96	0.92
Grain drill w/press wheels												
<2.4 m drill width	6.5–11.0	8.0	55–80	70	rows	400	0.0	0.0	1.0	1.0	1.0	1.0
2.4–3.7 m drill width	6.5–11.0	8.0	55–80	70	rows	300	0.0	0.0	1.0	1.0	1.0	1.0
>3.7 m drill width	6.5–11.0	8.0	55–80	70	rows	200	0.0	0.0	1.0	1.0	1.0	1.0
Grain drill, no-till												
1 fluted coulter/row	6.5–11.0	8.0	55–80	70	rows	720	0.0	0.0	1.0	0.92	0.92	0.79
Pneumatic drill	6.5–11.0	9.0	50–75	65	m	3700	0.0	0.0	1.0	1.0	1.0	1.0

Table II Rotary Power Requirement Parameters

Implement	Field Speed S (km h⁻¹) Range	Mean Value	Field Efficiency E_f (%) Range	Mean Value	a (kW)	b (kW m⁻¹)	c (kWh t⁻¹)
Baler, small rectangular	4.0–10.0	6.5	60–85	75	2.0	0	1.0[1]
Baler, large rectangular bales	6.5–13.0	8.0	70–90	85	4.0	0	1.3
Baler, large round (variable chamber)	5.0–13.0	8.0	55–75	80	4.0	0	1.1
Baler, large round (fixed chamber)	5.0–13.0	8.0	55–75	65	2.5	0	1.8
Combine, small grains	3.0–6.5	5.0	65–80	70	20.0	0	3.6[2]
Combine, corn	3.0–6.5	4.0	60–75	65	35.0	0	1.6[2]
Cotton picker	3.0–6.0	4.5	60–75	70	0	9.3	0
Forage harvester, corn silage	2.5–10.0	5.5	60–85	65	6.0	0	3.3[3]
Forage harvester, direct-cut	2.5–10.0	5.5	60–85	65	6.0	0	5.7[3]
Mower–conditioner, disk	8.0–19.0	11.0	75–90	80	0	8	0
Mower, cutterbar	5.0–10.0	8.0	75–85	80	0	1.2	0
Mower, disk	8.0–19.0	11.0	75–90	80	0	5.0	0
Mower, flail	8.0–19.0	11.0	75–90	80	0	10.0	0
Mower–conditioner, cutterbar	5.0–10.0	8.0	75–85	80	0	4.5	0
Potato harvester[4]	2.5–6.5	4.0	55–70	60	0	10.7	0
Rake, side delivery	6.5–13.0	10.0	70–90	80	0	0.4	0
Rake, rotary	6.5–13.0	10.0	70–90	80	0	2.0	0
Tedder	6.5–13.0	10.0	70–90	80	0	1.5	0

[1] Increase by 20% for straw.

[2] Based upon material-other-than-grain, MOG, throughput for small grains and grain throughput for corn. For PTO driven machine, reduced parameter a by 10 kW.

[3] Throughput is units of dry matter per hour with a 9 mm length of cut. At a specific throughput, a 50% reduction in the length of cut setting or the use of a recutter screen increases power 25%.

[4] Total power requirement must include a draft of 11.6 kN m⁻¹ (±40%) for potato harvesters and 5.6 kN m⁻¹ (±40%) for beet harvesters. A row spacing of 0.86 m for potatoes and 0.71 m for beets is assumed.

References

1. Farmer BH. Perspectives on the 'green revolution' in south asia. *Mod Asian Stud* 1986;**20**:175.

2. Sturrock FG, Cathie J, Payne TA. Economies of scale in farm mechanization. A study of costs of large and small farms. *Agric Enterp Stud Engl Wales* 1977;**56**.

3. Witney B. *Choosing & using farm machines*. 1988.

4. Sørensen CG, et al. Conceptual model of a future farm management information system. *Comput Electron Agric* 2010;**72**:37–47.

5. Sorensen CG, et al. A user-centric approach for information modelling in arable farming. *Comput Electron Agric* 2010;**73**:44–55.

6. Busato P, Berruto R, Cornelissen R. Bioenergy farm Project: economic and energy analysis of biomass by web application. In: *19th European Biomass Conference*; 2011.

7. Van Elderen E. *Heuristic strategy for scheduling farm operations*. PUDOC; 1977.

8. Goense D, Blaauw SK. *Fundamentals of agricultural mechanisation systems*. The Netherlands: Dept. of Agricultural Engineering and Physics, Wageningen Agricultural University; 1996.

9. Adam EE, Ebert RJ. *Production and operations management: concepts, models, and behavior*. Prentice Hall; 1989.

10. Adam EE, Ebert RJ. *Production and operations management: concepts, models, and behavior*. Prentice Hall; 1992.

11. Svendsen GB. The influence of interface style on problem solving. *Int J Man Mach Stud* 1991;**35**: 379–97.

12. Heizer J, Render B. *Production and operations management: strategic and tactical decisions*. Prentice Hall; 1996.

13. Meredith JR. *The management of operations: a conceptual emphasis*. Wiley; 1992.

14. Hunt D. *Farm power and machinery management*. IOWA University Press; 2001. p. 77–93. in I.

15. Nielsen V, Sørensen CG. Technical farm management a program for calculation of work requirement. *Work Capacity, Work Budget, Work Profile* 1993.

16. Kroeze C. *Nitrous Oxide (N_2O)-emission inventory and options for control in The Netherlands*. 1994.

17. Nielsen V, Sørensen C. *Green fields-operational analysis and model simulation*. 1994.

18. Russell SJ, Norvig P. Artificial intelligence: a modern approach. *Neurocomputing* 1995;**9**.

19. Standards ASAE. In: *Engineering practices and data developed and adopted by the American society of agricultural engineers*. ASAE, MI, U; 1974. p. 293–4.

20. Pinedo M. *Scheduling – theory, algorithms and systems*. Prentice-Hall Inc.; 1995.

21. Gantt HL. *Work, wages, and profits*. The Engineering Magazine CO; 1913.

22. Graves SC. A review of production scheduling. *Oper Res* 1981;**29**:646–75.

23. Rinnooy K. *Machine scheduling problems*. Springer US; 1976.

24. Dreger J. *Project management effective scheduling*. Van Nostrand Reinhold; 1992.

25. Link DA, Bockhop CW. Mathematical approach to farm machine scheduling. *Trans ASAE* 1964;**7**:16.

26. Peart RM, Von Bargen K, Deason DL. Network analysis in agricultural systems engineering. *Trans ASAE* 1970;**13**:849–53.

27. Hestermann C, Wolber M. A comparison between operations research models and real world scheduling problems. In: *Proceedings of the European Conference on Intelligent Management Systems in Operations*. University of Salford; 1997.

28. Parker RG. *Deterministic scheduling theory*. Chapman & Hall; 1995.

29. Collinot A, Le Pape C, Pinoteau G. SONIA: a knowledge-based scheduling system. *Artif Intell Eng* 1988;**3**:86–94.

30. Taunton JC, Ready CM. Intelligent dynamic production scheduling. *Food Res Int* 1994;**27**:111–6.

31. Fox MS. Constraint-directed search: a case study of job-shop scheduling. *J Chem Inf Model* 1983;**207**. https://doi.org/10.1017/CBO9781107415324.004.

32. Smith SFA. Constraint-Based framework for reactive management of factory schedules. In: *Intelligent manufacturing*. Benjamin/Cummings; 1987. p. 113–30.

33. Sadeh N. Micro-opportunistic scheduling: the micro-boss factory scheduler. In: *Intelligent scheduling*; 1994. p. 99–137.

34. Staar MK. *Production management – systems and synthesis*. Prentice Hall, Inc.; 1972.

35. Monks JG. *Operations management, theory and problems*. McGraw-Hill Book Company; 1987.

36. Michelsen AU. *Production and economy. Introduction to operational management*, vol. 40; 1992.

37. Plant RT. Utilising formal specifications in the development of knowledge-based systems. *Artif Intell Softw Eng* 1991.

38. Turban E. *Decision support and expert systems: management support systems*. Englewood Cliffs, NJ: Prentice Hall; 1995.

39. Srivastava AK, Goering CE, Rohrbach RP. Engineering principles of agricultural machines. In: *ASAE textbook*, vol. 6; 1993.

40. Goense D, Hofstee JW. Communication between and within agricultural equipment. In: *Society of automotive engineers international off-highway and powerplant congress and exposition*, vol. 6; 1994.

41. Ushma V. *Production & operations management*. 2017.

42. F. M. A strategic selection model of forage conservation techniques for dairy farms. In: *XII CGIR world congress on agricultural engineering*; 1994. p. 1072–87.

43. Hartz O. *Produktion-integreret Teknisk styring*. 1994.

44. ASABE. *D497.6: agricultural machinery management data*. 2009.

45. Kelly A, Stentz A. Analysis of requirements for high speed rough terrain autonomous mobility. I. Throughput and response. In: *Proceedings of international conference on robotics and automation*, vol. 4. IEEE; 1997. p. 3318–25.

46. Dubins LE. On curves of minimal length with a constraint on average curvature, and with prescribed initial and terminal positions and tangents. *Am J Math* 1957;**79**:497.

47. Reeds JA, Shepp LA. Optimal paths for a car that goes both forwards and backwards. *Pacific J Math* 1990;**145**:367–93.

48. Triggs B. *Motion planning for nonholonomic vehicles: an introduction*. 1993.

49. Reid JF, Zhang Q, Noguchi N, Dickson M. Agricultural automatic guidance research in North America. *Comput Electron Agric* 2000;**25**:155–67.

50. Lechner W, Baumann S. Global navigation satellite systems. *Comput Electron Agric* 2000;**25**:67–85.

51. Lipiński AJ, Markowski P, Lipiński S, Pyra P. Precision of tractor operations with soil cultivation implements using manual and automatic steering modes. *Biosyst Eng* 2016;**145**:22–8.

52. Bochtis DD, Vougioukas SG. Minimising the non-working distance travelled by machines operating in a headland field pattern. *Biosyst Eng* 2008;**101**:1−12.

53. Laporte G. The traveling salesman problem: an overview of exact and approximate algorithms. *Eur J Oper Res* 1992;**59**:231−47.

54. Padberg M, Rinaldi G. A branch-and-cut algorithm for the resolution of large-scale symmetric traveling salesman problems. *SIAM Rev* 1991;**33**:60−100.

55. Crowder H, Padberg MW. Solving large-scale symmetric travelling salesman problems to optimality. *Manag Sci* 1980;**26**:495−509.

56. Grötschel M, Holland O. Solution of large-scale symmetric travelling salesman problems. *Math Program* 1991;**51**:141−202.

57. Rosenkrantz DJ, Stearns RE, Lewis PM. An analysis of several heuristics for the traveling salesman problem. In: *Fundamental problems in computing*. Netherlands: Springer; 2009. p. 45−69. https://doi.org/10.1007/978-1-4020-9688-4_3.

58. Reinelt G. *The traveling salesman: computational solutions for TSP applications*. Springer-Verlag; 1994.

59. Dijkstra EW. A note on two problems in connexion with graphs. *Numer Math* 1959;**1**:269−71.

60. Hart P, Nilsson N, Raphael B. A formal basis for the heuristic determination of minimum cost paths. *IEEE Trans Syst Sci Cybern* 1968;**4**:100−7.

61. Clarke G, Wright JW. Scheduling of vehicles from a central depot to a number of delivery points. *Oper Res* 1964;**12**:568−81.

62. Segerstedt A. A simple heuristic for vehicle routing − a variant of Clarke and Wright's saving method. *Int J Prod Econ* 2014;**157**:74−9.

63. Lin S, Kernighan BW. An effective heuristic algorithm for the traveling-salesman problem. *Oper Res* 1973;**21**:498−516.

64. Potvin JY, Rousseau JM. An exchange heuristic for routing problems with time windows. *J Oper Res Soc* 1995;**46**:1433−46.

65. Gunnarsson C, Vågström L, Hansson P-A. Logistics for forage harvest to biogas production—timeliness, capacities and costs in a Swedish case study. *Biomass Bioenergy* 2008;**32**:1263−73.

66. Landers A. *Farm machinery: selection, investment and management*. Farming Press; 2000.

67. Bochtis DD, Sørensen CG, Green O, Moshou D, Olesen J. Effect of controlled traffic on field efficiency. *Biosyst Eng* 2010;**106**:14−25.

68. Mayer RJ, et al. *IDEF3 process description capture method report*. 1995.

69. Pavlou D, et al. Functional modeling for green biomass supply chains. *Comput Electron Agric* 2016;**122**:29−40.

70. Busato P, Berruto R. Minimising manpower in rice harvesting and transportation operations. *Biosyst Eng* 2016;**151**:435−45.

71. ASAE. ASAE EP496.3-agricultural machinery management. *Am Soc Agric Biol Eng* 2009;**3**:354−7.

72. American Society of Agricultural, Biological Engineers. ASAE D497.7 MAR2011 agricultural machinery management data. *Test* 2011;**9**.

73. Bowers W. *Machinery management*. J Deere Co. Service Publications; 1992.

74. ASAE D497.4. Agricultural machinery management data. *Am Soc Agric Biol Eng* 2005;**4**.

75. Stockdale EA, et al. Agronomic and environmental implications of organic farming systems. *Adv Agron* 2001;**70**:261−327.

76. Chapman DF, Kenny SN, Beca D, Johnson IR. Pasture and forage crop systems for non-irrigated dairy farms in southern Australia. 1. Physical production and economic performance. *Agric Syst* 2008;**97**:108–25.

77. de Toro A. Assessment of field machinery performance in variable weather conditions using discrete event simulation. *Acta Univ Agric Sueciae Agricar* 2004.

78. Witney BD, Eradat Oskoui K. The basis of tractor power selection on arable farms. *J Agric Eng Res* 1982;**27**:513–27.

79. Witney BD. *Choosing and using farm machines*. Land Technology Ltd; 1995.

80. Droogers P, Fermont A, Bouma J. Effects of ecological soil management on workability and trafficability of a loamy soil in The Netherlands. *Geoderma* 1996;**73**:131–45.

81. Munkholm LJ. Soil friability: a review of the concept, assessment and effects of soil properties and management. *Geoderma* 2011;**167–168**:236–46.

82. Mosaddeghi MR, Morshedizad M, Mahboubi AA, Dexter AR, Schulin R. Laboratory evaluation of a model for soil crumbling for prediction of the optimum soil water content for tillage. *Soil Tillage Res* 2009;**105**:242–50.

83. Rounsevell MDA, Jones RJA. A soil and agroclimatic model for estimating machinery work-days: the basic model and climatic sensitivity. *Soil Tillage Res* 1993;**26**:179–91.

84. Sørensen CG. Workability and machinery sizing for combine harvesting. *Agric Eng Int CIGR J* 2003.

85. Mygdakos E, Gemtos TA. IT—information technology and the human interface: reliability of cotton pickers and its effect on harvesting cost. *Biosyst Eng* 2002;**82**:381–91.

86. Robert A, Tufts RA. Failure frequency and downtime duration effects on equipment availability. *Trans ASAE* 1985;**28**:999–1002.

87. Sørensen CG, Nielsen V. Operational analyses and model comparison of machinery systems for reduced tillage. *Biosyst Eng* 2005;**92**:143–55.

88. ASAE D497.5. Agricultural machinery management data. In: *Asabe STANDARD 2009*, vol. I; 2009. p. 360–7.

89. Suh J. Theory and reality of integrated rice–duck farming in Asian developing countries: a systematic review and SWOT analysis. *Agric Syst* 2014;**125**:74–81.

90. Phadermrod B, Crowder RM, Wills G. Importance-Performance Analysis based SWOT analysis. *Int J Inf Manage* 2016. https://doi.org/10.1016/J.IJINFOMGT.2016.03.009.

91. Bochtis DD, Sørensen CGC, Busato P. Advances in agricultural machinery management: a review. *Biosyst Eng* 2014;**126**:69–81.

92. Kolisch R, Padman R. An integrated survey of deterministic project scheduling. *Omega* 2001;**29**:249–72.

93. Möhring RH. In: *Scheduling problems with a singular solution*; 1982. p. 225–39. https://doi.org/10.1016/S0304-0208(08)72454-X.

94. Möhring RH, Radermacher FJ. The order-theoretic approach to scheduling: the deterministic case. In: *Advances in project scheduling*. Elsevier; 1989. p. 29–66. https://doi.org/10.1016/B978-0-444-87358-3.50006-7.

95. Möhring RH, Radermacher FJ. The order-theoretic approach to scheduling: the stochastic case. In: *Advances in project scheduling*. Elsevier; 1989. p. 497–531. https://doi.org/10.1016/B978-0-444-87358-3.50024-9.

96. Garey MR, Johnson DS, Sethi R. The complexity of flowshop and jobshop scheduling. *Math Oper Res* 1976;**1**:117–29.

97. Jain AS, Meeran S. Deterministic job-shop scheduling: past, present and future. *Eur J Oper Res* 1999;**113**:390–434.

98. Meira LAA, Martins PS, Menzori M, Zeni GA. *Multi-objective vehicle routing problem applied to large scale post office deliveries.* 2017.

99. Brucker P, Jurisch B, Sievers B. A branch and bound algorithm for the job-shop scheduling problem. *Discret Appl Math* 1994;**49**:107–27.

100. Adams J, Balas E, Zawack D. The shifting bottleneck procedure for job shop scheduling. *Manage Sci* 1988;**34**:391–401.

101. van Laarhoven PJM, Aarts EHL, Lenstra JK. Job shop scheduling by simulated annealing. *Oper Res* 1992;**40**:113–25.

102. Dell'Amico M, Trubian M. Applying tabu search to the job-shop scheduling problem. *Ann Oper Res* 1993;**41**:231–52.

103. Cheng R, Gen M, Tsujimura Y. A tutorial survey of job-shop scheduling problems using genetic algorithms, part II: hybrid genetic search strategies. *Comput Ind Eng* 1999;**36**:343–64.

104. Blum C, Sampels M. An ant colony optimization algorithm for shop scheduling problems. *J Math Model Algorithms* 2004;**3**:285–308.

105. Abdullah S, Aickelin U, Burke E, Mohamed Din A, Qu R. Investigating a hybrid metaheuristic for job shop rescheduling. In: *Progress in artificial life.* Berlin Heidelberg: Springer; 2007. p. 357–68. https://doi.org/10.1007/978-3-540-76931-6_31.

106. Dorndorf U, Pesch E, Phan-Huy T. Constraint propagation and problem decomposition: a pre-processing procedure for the job shop problem. *Ann Oper Res* 2002;**115**:125–45.

107. Bräysy O, Gendreau M. Tabu search heuristics for the vehicle routing problem with time windows. *Top* 2002;**10**:211–37.

108. Cordeau JF, Gendreau M, Laporte G, Potvin JY, Semet F. A guide to vehicle routing heuristics. *J Oper Res Soc* 2002;**53**:512–22.

109. Laporte G. The vehicle routing problem: an overview of exact and approximate algorithms. *Eur J Oper Res* 1992;**59**:345–58.

110. Osman IH. Meta strategy simulated annealing and tabu search algorithms for the vehicle routing problem. *Ann Oper Res* 1993;**41**:421–51.

111. Jia H, Li Y, Dong B, Ya H. An improved tabu search approach to vehicle routing problem. *Procedia Soc Behav Sci* 2013;**96**:1208–17.

112. Bell JE, McMullen PR. Ant colony optimization techniques for the vehicle routing problem. *Adv Eng Inform* 2004;**18**:41–8.

113. Rousseau LM, Gendreau M, Pesant G, Focacci F. Solving VRPTWs with constraint programming based column generation. *Ann Oper Res* 2004;**130**:199–216.

114. Zhang W, Dietterich T. A reinforcement learning approach to job-shop scheduling. *Int Jt Conf Artif* 1995:1114–20.

115. Zomaya AY, Clements M, Olariu S. A framework for reinforcement-based scheduling in parallel processor systems. *IEEE Trans Parallel Distrib Syst* 1998;**9**:249–60.

116. Bochtis DD. Machinery management in bio-production systems: planning and scheduling aspects. *Agric Eng Int CIGR J* 2010;**12**:55–63.

117. Dantzig G, Fulkerson R, Johnson S. Solution of a large-scale traveling-salesman problem. *J Oper Res Soc Am* 1954;**2**:393–410.

118. Orman AJ, Williams HPA. Survey of different integer programming formulations of the travelling salesman problem. In: *Optimisation, econometric and financial analysis.* Berlin Heidelberg: Springer; 2007. p. 91–104. https://doi.org/10.1007/3-540-36626-1_5.

119. Gavish B, Graves SC. *The travelling salesman problem and related problems.* 1978.

120. Öncan T, Altınel K, Laporte G. A comparative analysis of several asymmetric traveling salesman problem formulations. *Comput Oper Res* 2009;**36**:637–54.

121. Brumitt, B.L, Stentz, A. GRAMMPS: a generalized mission planner for multiple mobile robots in unstructured environments. in Proceedings 1998 IEEE International Conference on Robotics and Automation (Cat. No.98CH36146) vol. 2, 1564–1571 (IEEE).

122. Brumitt, B.L, Stentz, A. Dynamic mission planning for multiple mobile robots. in Proceedings of IEEE International Conference on Robotics and Automation vol. 3, 2396–2401 (IEEE).

123. Angel RD, Caudle WL, Noonan R, Whinston A. Computer-assisted school bus scheduling. *Manage Sci* 1972;**18**. B-279–B-288.

124. Svestka JA, Huckfeldt VE. Computational experience with an *M*-salesman traveling salesman algorithm. *Manage Sci* 1973;**19**:790–9.

125. Dantzig GB, Ramser JH. The truck dispatching problem. *Manage Sci* 1959;**6**:80–91.

126. Diestel R. *Graph theory*. Springer; 1990.

127. Ricks D, Woods T, Sterns J. Chain management and marketing performance in fruit industry. *Acta Hortic* 2002;**536**:661–8.

128. van der Vorst JGAJ. In: *Performance levels in food traceability and the impact on chain design: Results of an international benchmark study*; 2004. p. 175–83.

129. van der Vorst JGAJ, Beulens AJM, van Dijk SJ. In: *Modelling and simulating SCM scenarios in food supply chains*; 2000. p. 419–28.

130. Tsolakis NK, Keramydas CA, Toka AK, Aidonis DA, Iakovou ET. Agrifood supply chain management: a comprehensive hierarchical decision-making framework and a critical taxonomy. *Biosyst Eng* 2014;**120**:47–64.

131. Iakovou E, Bochtis D, Vlachos D, Aidonis D. Sustainable agrifood supply chain management. *Supply Chain Manag Sustain Food Netw* 2015. https://doi.org/10.1002/9781118937495.ch1.

132. Dekker R, Bloemhof-Ruwaard J, Mallidis I. A hierarchical decision-making framework for quantitative green supply chain management. In: *Supply chain management for sustainable food networks*. John Wiley & Sons Ltd; 2016. p. 129–57. https://doi.org/10.1002/9781118937495.ch5.

133. Bovea MD, Pérez-Belis V. A taxonomy of ecodesign tools for integrating environmental requirements into the product design process. *J Clean Prod* 2012;**20**:61–71.

134. Philips RE, Blevins RL, Thomas GW, Foge WW, Phillips SH. No-tillage agriculture. *Science* 1980;**208**:1108–13.

135. Conforti P, Giampietro M. Fossil energy use in agriculture: an international comparison. *Agric Ecosyst Environ* 1997;**65**:231–43.

136. Refsgaard K, Halberg N, Kristensenb ES. Energy utilization in crop and dairy production in organic and conventional livestock production systems. *Agric Syst* 1998;**57**:599–630.

137. Dalgaard T, Porter JR, Halberg N. A model for fossil energy use in Danish agriculture used to compare organic and conventional farming. *Agric Ecosyst Environ* 2001;**87**:51–65.

138. Moerschner J, Gerowitt B. Direct and indirect energy use in arable farming e an example on winter wheat in Northern Germany. *Agric Data for Life Cycle Assess* 2000;**1**.

139. Dalgaard R, Halberg N, Kristensen IS, Larsen I. Modelling representative and coherent Danish farm types based on farm accountancy data for use in environmental assessments. *Agric Ecosyst Environ* 2006;**117**:223–37.

140. Reicosky DC, Dugas WA, Torbert HA. Tillage induced soil carbon dioxide loss from different cropping systems. *Soil Tillage Res* 1997;**41**:105–18.

141. Franzluebbers AJ, Langdale GW, Schomberg HH. Soil carbon, nitrogen, and aggregation in response to type and frequency of tillage. *Soil Sci Soc Am J* 1999;**63**:349.

142. Schlesinger GH. Changes in soil carbon storage and associated properties with disturbance and recovery. In: Trabella JR, et al., editors. *The changing carbon cycle: a global analysis.* Springer Verlag; 1985. p. 194–220.

143. Mutegi JK. *Soil carbon sequestration and nitrous oxide emission as affected by tillage, cover crops and crop rotations.* Aarhus University; 2011.

144. Chatskikh D, Olesen JE, Hansen EM, Elsgaard L, Petersen BM. Effects of reduced tillage on net greenhouse gas fluxes from loamy sand soil under winter crops in Denmark. *Agric Ecosyst Environ* 2008;**128**:117–26.

145. Sørensen CG, Nielsen V. Operational analyses and model comparison of machinery systems for reduced tillage. *Biosyst Eng* 2005;**92**:143–55.

146. KTBL. *Data for farm calculations in agriculture.* 1994.

147. Bowers W. Agricultural field equipment. *Energy farm Prod* 1992;**6**:117–29.

148. Stout BA. Handbook of energy for world agriculture. *Agric Syst* 1990;**35**.

149. Kalk WD, Hulsbergen KJ. Method for considering the materialized energy (indirect energy consumption) in capital goods on energy balance sheets of farms. *Kuhn-archiv* 1996;**90**:41–56.

150. Kranzlein T. *Forecasting energy use in agriculture.* 2004.

151. Doering OC. Accounting for energy in farm machinery and buildings. In: *Handbook of energy utilization in agriculture.* CRC Press; 1980. p. 9–14.

152. Nagy CN. *Energy and greenhouse gas coefficients inputs used in agriculture.* 2000.

153. Dyer JA, Desjardins RL. Carbon dioxide emissions associated with the manufacturing of tractors and farm machinery in Canada. *Biosyst Eng* 2006;**93**:107–18.

154. Stripple H. *Life cycle assessment of road. A pilot study for inventory analysis.* IVL Swedish Environmental Research Institute Ltd; 2001.

155. Althaus HJ, Classen M, Blaser S, Jungbluth DN. *Life cycle inventories of metals data v1.1.* 2004.

156. Djekic I. New approach to life cycle analysis of self-propelled agricultural machines. *Agric Eng Int CIGR E J* 2006;**VIII**.

157. Kellenberger D, Althaus H-J, Künniger T, Jungbluth DN. *Life cycle inventories of building products data.* 2004.

158. Heller MC, Keoleian GA, Volk TA. Life cycle assessment of a willow bioenergy cropping system. *Biomass Bioenergy* 2003;**25**:147–65.

159. Cho J-G, Han M. *LCA applications in LG for the cleaner production.* 2001.

160. Audsley E. *Harmonisation of environmental life cycle assessment for agriculture.* 1997.

161. Nemecek T, Erzinger S. Modelling representative life cycle inventories for Swiss arable crops. *Int J Life Cycle Assess* 2005;**10**:68–76.

162. Laursen B. *Machinery costs in relation to machine age and yearly use.* 1993.

163. Agrimach. *Agrimach multimedia. The International source for machinery.* 2004. Available at: http://www.agrimach.com/en/.

164. Nielsen V, Sørensen CG. *'DRIFT'. A program for estimating labour requirement, labour capacity, labour budget and labour profile.* 1993.

165. Sørensen CG, Jacobsen BH, Sommer SG. An assessment tool applied to manure management systems using innovative technologies. *Biosyst Eng* 2003;**86**:315–25.

166. ISO. *Environmental management—life cycle assessment—life cycle impact assessment.* 2000.

167. SETAC-Europe. Best available practice regarding impact categories and category indicators in life cycle impact assessment. *Int J Life Cycle Assess* 1999;**4**:167–74.

168. Manzano-Agugliaro F, Cañero-Leon R. Economics and environmental analysis of mediterranean greenhouse crops. *Afr J Agri Res* 2010;**5**:3009—16.

169. Bechar A, Eben-Chaime M. Hand-held computers to increase accuracy and productivity in agricultural work study. *Int J Product Perform Manag* 2014;**63**:194—208.

170. Tozer PR. Uncertainty and investment in precision agriculture — is it worth the money? *Agric Syst* 2009;**100**:80—7.

171. Bisgaard M, Vinther D, Ostergaard KZ. Modelling and fault-tolerant control of an autonomous wheeled robot. *Group* 2004.

172. Van Ota T, Bontsema J, Hayashi S, Kubota K, Van Henten EJ, Van Os EA, et al. Development of a cucumber leaf picking device for greenhouse production. *Biosyst Eng* 2007;**98**:381—90.

173. Van Henten EJ, Schenk EJ, van Willigenburg LG, Meuleman J, Barreiro P. Collision-free inverse kinematics of the redundant seven-link manipulator used in a cucumber picking robot. *Biosyst Eng* 2010;**106**:112—24.

174. Huang L, Yang SX, He D. Abscission point extraction for ripe tomato harvesting robots. *Intell Autom Soft Comput* 2012;**18**:751—63.

175. Zhao Y, Gong L, Huang Y, Liu C. Robust tomato recognition for robotic harvesting using feature images fusion. *Sensors* 2016;**16**.

176. Bac CW, Hemming J, Van Henten EJ. Stem localization of sweet-pepper plants using the support wire as a visual cue. *Comput Electron Agric* 2014;**105**:111—20.

177. Schor N, et al. Robotic disease detection in greenhouses: combined detection of powdery mildew and tomato spotted wilt virus. *IEEE Robot Autom Lett* 2016;**1**:354—60.

178. Vitzrabin E, Edan Y. Changing task objectives for improved sweet pepper detection for robotic harvesting. *IEEE Robot Autom Lett* 2016;**1**:578—84.

179. Cui Y, Gejima Y, Kobayashi T, Hiyoshi K, Nagata M. Study on cartesian-type strawberry-harvesting robot. *Sens Lett* 2013;**11**:163—8.

180. Hayashi S, et al. Evaluation of a strawberry-harvesting robot in a field test. *Biosyst Eng* 2010;**105**:160—71.

181. Xu Y, Imou K, Kaizu Y, Saga K. Two-stage approach for detecting slightly overlapping strawberries using HOG descriptor. *Biosyst Eng* 2013;**115**:144—53.

182. Blanes C, Ortiz C, Mellado M, Beltrán P. Assessment of eggplant firmness with accelerometers on a pneumatic robot gripper. *Comput Electron Agric* 2015;**113**:44—50.

183. Mann MP, Rubinstein D, Shmulevich I, Linker R, Zion B. Motion planning of a mobile cartesian manipulator for optimal harvesting of 2-D crops. *Trans ASABE* 2014;**57**:283—95.

184. Mann M, Zion B, Shmulevich I, Rubinstein D. Determination of robotic melon harvesting efficiency: a probabilistic approach. *Int J Prod Res* 2016;**54**:3216—28.

185. Figliolini G, Rea P. Overall design of Ca.U.M.Ha. robotic hand for harvesting horticulture products. *Robotica* 2006;**24**:329—31.

186. Irie N, Taguchi N, Horie T, Ishimatsu T. Asparagus harvesting robot coordinated with 3-D vision sensor. In: *Proceedings of the IEEE international conference on industrial technology*; 2009. https://doi.org/10.1109/ICIT.2009.4939556.

187. Choi KH, et al. Morphology-based guidance line extraction for an autonomous weeding robot in paddy fields. *Comput Electron Agric* 2015;**113**:266—74.

188. Tamaki K, et al. A robot system for paddy field farming in Japan. In: *IFAC proceedings volumes (IFAC-PapersOnline)*, vol. 4; 2013. p. 143—7.

189. Reed JN, Miles JS, Butler J, Baldwin M, Noble R. Automation and emerging technologies for automatic mushroom harvester development. *J Agric Eng Res* 2001;**78**:15—23.

190. Tanigaki K, Fujiura T, Akase A, Imagawa J. Cherry harvesting robot. *Comput Electron Agric* 2008;**63**: 65−72.

191. Baeten J, Donné K, Boedrij S, Beckers W, Claesen E. Autonomous fruit picking machine: a robotic apple harvester. In: *Springer tracts in advanced robotics*, vol. 42; 2008. p. 531−9.

192. Nguyen TT, Kayacan E, De Baedemaeker J, Saeys W. Task and motion planning for apple harvesting robot. In: *IFAC proceedings volumes (IFAC-PapersOnLine)*, vol. 4; 2013. p. 247−52.

193. Brown GK. New mechanical harvesters for the Florida citrus juice industry. In: *HortTechnology*, vol. 15; 2005. p. 69−72.

194. Hannan MW, Burks TF, Bulanon DMA. Real-time machine vision algorithm for robotic citrus harvesting written for presentation at the 2007 ASABE annual international meeting sponsored by ASABE. *Am Soc Agric Biol Eng Annu Int Meet* 2007;**300**:1−12.

195. Yoon B, Kim S. Design of paddy weeding robot. In: *44th International symposium on robotics, ISR 2013*; 2013. https://doi.org/10.1109/ISR.2013.6695740.

196. Hiremath S, Van Der Heijden G, Van Evert FK, Stein A. The role of textures to improve the detection accuracy of Rumex obtusifolius in robotic systems. *Weed Res* 2012;**52**:430−40.

197. Ogawa Y, Kondo N, Monta M, Shibusawa S. Spraying robot for grape production. *Springer Tracts Adv Robot* 2006;**24**:539−48.

198. Slaughter DC, Giles DK, Downey D. Autonomous robotic weed control systems: a review. *Comput Electron Agric* 2008;**61**:63−78.

199. Van Evert FK, Samsom J, Polder G, Vijn M, Van Dooren HJ, Lamaker A, et al. A robot to detect and control broadleaved dock (*Rumex obtusifolius* L.) in grassland. *J F Robot* 2011;**28**:264−77.

200. Hu J, Yan X, Ma J, Qi C, Francis K, Mao H. Dimensional synthesis and kinematics simulation of a high speed plug seedling transplanting robot. *Comput Electron Agric* 2014;**107**:64−72.

201. Haibo L, Shuliang D, Zunmin L, Chuijie Y. Study and experiment on a wheat precision seeding robot. *J Robot* 2015;**2015**.

202. Mao H, Han L, Hu J, Kumi F. Development of a pincette-type pick-up device for automatic transplanting of greenhouse seedlings. *Appl Eng Agric* 2014;**30**:547−56.

203. Nagasaka Y, Mizushima A, Noguchi N, Saito H, Kubayashi K. Unmanned rice-transplanting operation using a GPS-guided rice transplanter with Long Mat-Type hydroponic seedlings. *Agric Eng Int CIGR Ejournal* 2007;**IX**.

204. Nagasaka Y, Taniwaki K, Otani R, Shigeta K. An automated rice transplanter with RTKGPS and FOG. *Agric Eng Int CIGR Ejournal* 2002;**IV**.

205. Chamen T, et al. Prevention strategies for field traffic-induced subsoil compaction: a review: Part 2. Equipment and field practices. *Soil Tillage Res* 2003;**73**:161−74.

206. Hamza MA, Anderson WK. Soil compaction in cropping systems: a review of the nature, causes and possible solutions. *Soil Tillage Res* 2005;**82**:121−45.

207. Bochtis DD, Sørensen CG, Jørgensen RN, Green O. Modelling of material handling operations using controlled traffic. *Biosyst Eng* 2009;**103**:397−408.

208. Chamen WCT, Audsley E. A study of the comparative economics of conventional and zero traffic systems for arable crops. *Soil Tillage Res* 1993;**25**:369−96.

209. Batte MT, Ehsani MR. The economics of precision guidance with auto-boom control for farmer-owned agricultural sprayers. *Comput Electron Agric* 2006;**53**:28−44.

210. Tullberg J. Tillage, traffic and sustainability—a challenge for ISTRO. *Soil Tillage Res* 2010;**111**: 26−32.

211. Taylor JH. Benefits of permanent traffic lanes in a controlled traffic crop production system. *Soil Tillage Res* 1983;**3**:385−95.

212. Tullberg JN, Yule DF, McGarry D. Controlled traffic farming—from research to adoption in Australia. *Soil Tillage Res* 2007;**97**:272–81.

213. Tullberg JN, Ziebarth PJ, Li Y. Tillage and traffic effects on runoff. *Aust J Soil Res* 2001;**39**:249.

214. Douglas JT, Crawford CE, Campbell DJ. Traffic systems and soil aerator effects on grassland for silage production. *J Agric Eng Res* 1995;**60**:261–70.

215. McPhee JE, Braunack MV, Garside AL, Reid DJ, Hilton DJ. Controlled traffic for irrigated double cropping in a semi-arid tropical environment: Part 2, tillage operations and energy use. *J Agric Eng Res* 1995;**60**:183–9.

216. Pierce FJ, Nowak P. Aspects of precision agriculture. *Adv Agron* 1999;**67**:1–85.

217. Fountas S, Aggelopoulou K, Gemtos TA. Precision agriculture: crop management for improved productivity and reduced environmental impact or improved sustainability. In: *Supply chain management for sustainable food networks*. John Wiley & Sons, Ltd; 2015. p. 41–65. https://doi.org/10.1002/9781118937495.ch2.

218. Pierpaoli E, Carli G, Pignatti E, Canavari M. Drivers of precision agriculture technologies adoption: a literature review. *Procedia Technol* 2013;**8**:61–9.

219. Bechar A, Vigneault C. Agricultural robots for field operations: concepts and components. *Biosyst Eng* 2016;**149**:94–111.

220. Bechar A, Vigneault C. Agricultural robots for field operations. Part 2: operations and systems. *Biosyst Eng* 2017;**153**:110–28.

221. Kutter T, Tiemann S, Siebert R, Fountas S. The role of communication and co-operation in the adoption of precision farming. *Precis Agric* 2011;**12**:2–17.

222. Knapper A, Van N,N, Christoph M, Hagenzieker M, Brookhuis K. The use of navigation systems in naturalistic driving. *Traffic Inj Prev* 2016;**17**:264–70.

223. Bryden KJ, Charlton JL, Oxley JA, Lowndes GJ. Acceptance of navigation systems by older drivers. *Gerontechnology* 2014;**13**:21–8.

224. Joubert JW, Meintjes S. Repeatability & reproducibility: implications of using GPS data for freight activity chains. *Transp Res Part B Methodol* 2015;**76**:81–92.

225. Bohte W, Maat K. Deriving and validating trip purposes and travel modes for multi-day GPS-based travel surveys: a large-scale application in The Netherlands. *Transp Res Part C Emerg Technol* 2009;**17**:285–97.

226. Halberg N. Indicators of resource use and environmental impact for use in a decision aid for Danish livestock farmers. *Agric Ecosyst Environ* 2001;**76**:17–30.

227. Sørensen CG, et al. Conceptual model of a future farm management information system. *Comput Electron Agric* 2010;**72**:37–47.

228. McCown RL. Changing systems for supporting farmers' decisions: problems, paradigms, and prospects. *Agric Syst* 2002;**74**:179–220.

229. Lewis T. Evolution of farm management information systems. *Comput Electron Agric* 1998;**19**:233–48.

230. Folinas D. A conceptual framework for business intelligence based on activities monitoring systems. *Int J Intell. Enterp* 2007;**1**:65.

231. O'Brien J. *Management information systems – managing information technology in the internetworked enterprise*. Irwin McGraw-Hill; 1999.

Index

'Note: Page numbers followed by "f" indicate figures and "t" indicate tables.'

A

Ackermann steering models, 52, 52f
Agricultural holding, 119
  activities for, 12–13
  distribution of, 13, 13f
Agricultural machinery
  advances and future trends in
    controlled-traffic farming (CTF), 200
    farm management information systems, 205–208
    precision farming management, 200–204
    robotics, 197–199
    satellite navigation, 204–205
  area capacity, 48–49, 50t
  direct cost
    approximate cost estimation method, 103–105
    operating cost, 90–103
    ownership cost, 79–90
  field efficiency
    ASABE Standards, 49–50, 51t
    basic routing optimization problem, 67–73
    fieldwork patterns, 51–56
    in-field nonworking traveled distance, 67
    modeling field operations, 65
    modeling headland turning, 57–63
    multiple machines, 66
    parallel tracks coverage system, 63–64
    part-time work elements, 50, 52f
  indirect cost
    reliability, 112–115
    timeliness, 105–108
    trafficability, 109–112
    workability, 108–109
  material capacity, 49
  part processes, 47–48
  productivity, 73–78
  total labor requirement, 47–48
  work components, 47, 47t
Agricultural production systems
  advantages, 1, 2f
  agricultural holding
    activities for, 12–13
    distribution of, 13, 13f
  agricultural machines, 17, 17f
  automation phase, 10–12
  disadvantages, 1, 2f
  industrial production
    customers, 16
    information, 16
    raw materials, 16
  information and communication technologies, 10–12, 11t
  mechanization phase
    benefits, 2
    economies of scales, 2–3, 3f
  phases, time period of, 1, 2f
  production process, 14f
    conversion process, 14
    inspection, 15
    manipulation, 15
    processing, 15
    storage, 15
    transformation process, 14
    transport, 15
  transformation processes, 17, 18t
  work organization phase, 7f
    advanced analytic tools, 9
    combined effect of, 6–8, 7f
    estimate effective field capacity, 5, 6f
    field efficiency, 5
    modern devices for, 5
    separation yield losses, 8, 9f

Agricultural production systems (*Continued*)
    task time, 5
    Taylorism, 4
    total operational cost, 8, 8f
    turning time, 5, 6f
Agricultural vehicle routing. *See* Vehicle
    routing problem (VRP)
Agriproducts supply chain operations
  biomass supply chain
    advanced planning system for, 184–185,
      184f
    harvesting and handling operational
      planning, 185
    on-line planning and execution control,
      185–186, 186f
    tactical planning, 185
    tangible assets, 185
  logistics
    agrofood and biomass logistics, 180
    challenges, 179–180
    concept of, 179, 179f
    defined, 179
    objectives, 179–180
  management
    decision-making process, 182, 183f
    flow types, 181, 182f
    interactions, 181, 182f
    stakeholders, 181, 182f
Algorithms
  approximate algorithms, 69
  bounding, 70
  branching, 69–70, 70f
  Clarke-Wright algorithm, 71–72
  combinatorial optimization, 68
  NP-hard, 67–68
  randomized search, 69
Approximate cost estimation method,
    103–105
Arable farm operations, 39, 40t
Artificial intelligence (AI), 27
Asymmetric TSP (aTSP), 169

B
Basic routing optimization problem
  algorithms
    approximate algorithms, 69

    bounding, 70
    branching, 69–70, 70f
    Clarke-Wright algorithm, 71–72
    combinatorial optimization, 68
    NP-hard, 67–68
    randomized search, 69
  field shape, 73, 74f, 75t
Biomass supply chain
  advanced planning system for, 184–185,
    184f
  harvesting and handling operational
    planning, 185
  on-line planning and execution control,
    185–186, 186f
  tactical planning, 185
  tangible assets, 185
Bounding, 70
Branching, 69–70, 70f

C
Capacity planning, 161, 168f
Capital consumption method, 88
Clarke-Wright algorithm
  heuristics
    Or-opt, 71–72, 72f
    R-opt, 71, 72f
    swap operations, 72, 73f
  randomization, 71
Controlled-traffic farming (CTF), 200,
    201f
Cooperative use of machinery, 156
  agricultural machinery rings, 155–156
  with common holdings, 155
  with independent holdings, 154–155
Critical path method (CPM), 31–32
CTF. *See* Controlled-traffic farming (CTF)

D
Danish cropping systems, 187–188
Declining balance method, 82–83
Depreciation
  declining balance method, 82–83
  estimated value method, 83
  fixed percentage method, 82–83
  resale value method, 83

sinking fund method, 81–82
straight-line method, 79–80
sum-of-years digits method, 80–81
Direct energy estimations
average fuel consumption, 188–189
operations scenarios, 188–189
Draft parameters, 210t–211t
Dubins paths, 53

E

Embodied energy, 189–190
Energy usage, 187–188
Engineering management
agricultural operations, 38, 38t, 40t
agricultural production systems, 37
controlling production systems
decision-making process, 39–41
management cycle, 42
managing agricultural operations, 42f,
43–45
model for, 39, 41f
describing and defining operations
management activities, 38–39, 38t
designing and organizing production
systems, 34–39
evolution of, 19, 19f
industrial production systems
customers, 37
information, 37
raw materials, 37
job management, 30–34
operations system's elements,
19–20, 20f
planning levels
artificial intelligence (AI), 27
decision-making, 23
defined, 21–22
demands, 24
evaluation, 22, 27–28, 29t
execution, 22, 27–28, 29t
generation and execution, 22
implementation, 27–28, 29t
inherent uncertainty, 22
"job-shop", 24
operational, 22, 28t

optimization, 25–26
planning problem, 21–22
strategic and tactical planning functions,
22, 24, 25t
tactical, 22, 26
production process, 35f
inspection, 36
manipulation, 36
processing, 36
storage, 36
transport, 36
project management, 30–34
Enterprise resource planning (ERP), 206
Equipment selection
available time for field operations
actual performance, 131
disk harrow, 131
maximum power, 132
required power, 132
available tractor power
agricultural holding, 129
maximum width, 130
power efficiency, 129
price replacement, 130
tillage costs, 130
total time, 130
execution of field operations on time
operating costs, 136
repairs and maintenance costs, 137
seed-drilling machine, 137
minimum cost method
machine operation cost, 133
plowing cost, 135
processing cost, 134
purchase price, 133
time limitation and untimely field
operations
annual use of, 139
seed-drilling machine, 140
Estimated value method, 83

F

Farm management information systems,
205
concept of, 206–207, 207f

Farm management information systems
  (*Continued*)
  enhancement of, 206
  enterprise resource planning (ERP), 206
  management information systems (MIS),
    206
Field efficiency
  ASABE Standards, 49–50, 51t
  basic routing optimization problem
    algorithms, 67–72
    field shape, 73
  fieldwork patterns
    Ackermann steering models, 52, 52f
    headland turnings, 53–56
    minimum dynamic turning radius,
      52–53
    minimum turning radius value,
      52–53
    tricycle model, 52, 52f
  in-field nonworking traveled distance, 67
  modeling field operations, 65, 65f
  modeling headland turning, 57–63
  multiple machines, 66
  parallel tracks coverage system
    maneuvering degree, 64
    number of tracks, 64
    working width, 64
  part-time work elements, 50, 52f
Fixed percentage method, 82–83
Fleet management problem (FMP), 165
Fuels, 95–97, 95f–96f, 96t

G
Gantt charts, 30
GIS-based method, 185
Global Navigation Satellite System (GNSS),
  204
Globaluaya Navigatsionnaya Sputnikovaya
  Sistema (GLONASS), 204
Green biomass supply chains, 185–186,
  186f
Greenhouse gas (GHG) emissions, 187–188
Green revolution, 1

H
Headland turnings
  Dubins paths, 53
  minimum turning radius, 54–56, 55f–57f
  modeling of, 57–58
    minimum width of, 60–63, 62t–63t
    T-turn, 59–60, 59f
    Π-turn, 58, 58f
    Ω-turn, 58, 59f
  Reeds-Shepp paths, 53–54
  two-dimensional vehicle turnings, 54, 54f

I
IDEF0, 76–77, 76f–77f
IDEF3, 76–77, 77t, 78f
Indirect energy estimations
  machine weights, 190–192
  operation and task times, 192–193
  self-propelled machines and
    manufacturing, 190, 191t
Inflation
  coefficient, 84
  comparison of methods, 85–86, 86f, 86t
  inflation rate, 85
Inherent uncertainty, 22
In-vehicle navigation systems, 204

J
Job management, 30–34
Job-shop scheduling problem (JSSP),
  164–165

L
Labor cost
  fixed cost
    annual depreciation, 100
    capital interest, 100
    housing, 102
    insurance and taxes, 100
  hourly cost, 97, 100f
  60 kW diesel power tractor, 98–100
  operating cost
    fuel cost, 102

labor cost, 103
lubrication cost, 103
repair and maintenance, 102
operational and total annual cost, 97, 98f
ownership, 97, 98f
total cost distribution, 97, 99f, 100–103
tractor cost of, 100–103, 101t
wage rates, 97
Life-cycle assessment (LCA) methodology,
187–188
environmental profile, 196
framework, 193, 193f
functional unit, 194–195
generic example of, 194, 194f
goal and scope, 193–194
impact assessment phase, 196
impact categories, 195t, 196
interpretation phase, 196
inventory analysis workflow, 195–196, 195f
ISO 14,040:2006, 194
streamlined LCA, 193
Logistics
agrofood and biomass logistics, 180
challenges, 179–180
concept of, 179, 179f
defined, 179
objectives, 179–180
Lubricants, 97

M

Machinery management systems
cooperative use of machinery, 154–156
professional machinery, 151–153
renting/leasing of, 156–157
self-owned machinery, 149–151
state machinery, 157–158
Machinery replacement
cost analysis, 142–143
cumulative cost per unit area, 143, 145f
deterioration, 141
good policy, 142
good selling price, 141
inadequacy, 141
obsolescence, 141
second-hand machinery, 145–148

time of, 142
total operation hours, 143, 144t
Machinery system
equipment selection
available time for field operations,
131–132
available tractor power, 129–131
execution of field operations on time,
135–138
minimum cost method, 133–135
time limitation and untimely field
operations, 138–140
machinery replacement
cost analysis, 142–143
cumulative cost per unit area, 143, 145f
deterioration, 141
good policy, 142
good selling price, 141
inadequacy, 141
obsolescence, 141
second-hand machinery, 145–148
time of, 142
total operation hours, 143, 144t
tractor selection
factors, 117
minimum cost, 117–125
timely completion of operations,
125–128
Management information systems (MIS),
206
Multiobjective optimization problems,
160–161
Multiple TSP (mTSP), 170–171

N

Navigation System with Time and Ranging
Global Positioning System (NAVSTAR
GPS), 204
Non-deterministic polynomial-time hard,
67–68

O

Operating cost
fuels, 95–97, 95f–96f, 96t
labor cost

Operating cost (*Continued*)
    hourly cost, 97, 100f
    60 kW diesel power tractor, 98–100
    operational and total annual cost, 97, 98f
    ownership, 97, 98f
    total cost distribution, 97, 99f, 100–103
    tractor cost of, 100–103, 101t
    wage rates, 97
  lubricants, 97
  repair and maintenance cost
    boom-type spreader, 94
    cumulative cost, 94
    functions, 90
    parameters, 90, 91t–92t
    percentage of purchase price, 90,
      92f–93f
Operational planning, 22, 28t
Optimization problem, 160–161
Ownership cost
  capital consumption method, 88
  depreciation, 79–83
  housing, 89
  inflation, 83–86
  insurance and taxes, 89–90
  interest
    interest charge, 87
    yearly ownership cost, 86–87, 87f

P
Parallel tracks coverage system
  maneuvering degree, 64
  number of tracks, 64
  working width, 64
Performance evaluation, 175–178
  assessment, 178
  discrete events, 176
  feedback, 178
  identification, 178
  monitoring, 177
  ordinary events, 176
  response, 178
  urgent events, 175
  visualization, 178
PERT. *See* Program evaluation and review
    technique (PERT)

Planning problem, 21–22
Precision farming management
  concept of, 201–202, 202f
  cost of, 204
  criteria, 203
  defined, 201
Professional machinery
  annual cost equation, 152
  cost of, 151
  cost of seeding, 153, 153f–154f
Program evaluation and review technique
    (PERT), 31–32
Project management, 30–34
Project scheduling problems (PSPs),
    162–163

R
Reeds-Shepp paths, 53–54
Reference information model, 38–39
Registration, 15
Reliability
  breakdown possibility, 114, 114t
  machine downtime, 113–114
Remaining value coefficients, 83, 84t
Renting/leasing, of machinery, 156–157
Repair and maintenance cost
  boom-type spreader, 94
  cumulative cost, 94
  functions, 90
  parameters, 90, 91t–92t
  percentage of purchase price, 90, 92f–93f
Resale value method, 83
Robotics, 11
  autonomous machines, 197–198
  benefits of, 198
  categories of, 198, 199f
  challenges, 199
  field operations, 198–199
Rotary power requirement parameters,
    212t

S
Satellite navigation systems, 204–205
Scheduling
  feasibility of, 30

Gantt charts, 30
job-shop scheduling problem (JSSP), 164–165
makespan minimization, 163
modeling of
    classification, 31, 32t
    factory environment, 31
    planning tools, 31–32
project scheduling problems (PSPs), 162–163
short-term, 30
stochastic project scheduling, 163
time-cost trade-off problems, 163
types, 167–168, 174f
Scrap value, 79
Second-hand machinery
    advantage, 145–146
    vs. new grass balers, 146, 148t
    second-hand grass baler, 146, 147t
Self-owned machinery, 149–151
Short-term planning, 27
Short-term scheduling, 30
Sinking fund method, 81–82
Soil workability, 109
Straight-line method, 79–80
Strategic planning, 22, 25–26, 25t
Streamlined LCA, 193
Sum-of-years digits method, 80–81
SWOT analysis, 160, 160f
Symmetric TSP (sTSP), 169

T
Tactical planning, 22, 25t, 26, 185
Tangible assets, 185
Task time planning
    scheduling
        job-shop scheduling problem (JSSP), 164–165
        makespan minimization, 163
        project scheduling problems (PSPs), 162–163
        stochastic project scheduling, 163
        time-cost trade-off problems, 163
        types, 167–168, 174f

solution techniques
    heuristics, 165
    metaheuristics, 165
    vehicle routing problem (VRP), 166
Time-cost trade-off problems, 163
Timeliness
    cereal timeliness penalties, 105–106, 106f
    coefficients, 108, 109t
    crop yield, 105–106, 106f
    loss factors, 108
    yield losses, 107, 107f
Tractor selection
    factors, 117
    minimum cost
        corn, 123
        cotton, 123
        crops and area, 119, 119t
        energy requirements, 123, 124t
        final cost equation, 118
        fixed costs, 119
        grain, 123
        movement of pairs, 121–122
        necessary elements, 119
        required power, 124–125, 125t
        rolling resistance factor, 123
        task log, 120, 120t
        traction and power requirements, 120–121, 121t
        value replacements, 122
    timely completion of operations
        average hourly charges, 128
        charge annual cost, 125–126
        time limitation, 127
        untimely field operations, 127
Trade-in value, 79
Trafficability
    defined, 109–111
    out-of-time field operation costs, 112
    soil moisture, 111–112
Tramline farming system, 200. See also Controlled-traffic farming (CTF)
Transportation, 15
Traveling salesman problem (TSP), 169
Tricycle model, 52, 52f

V

Vehicle routing problem (VRP), 166
  area coverage planning, 171–172
  asymmetric TSP (aTSP), 170
  integer programming problem, 169
  multiple TSP (mTSP), 170–171
  for primary units
    farm and four fields, 173–174, 174f
    four machinery cases, 173, 173t
    operational time for, 174, 175f–177f

traditional and optimal planning,
  172–173
traveling salesman problem (TSP), 169

W

Workability
  defined, 108
  soil moisture conditions, 109, 110t
  soil workability, 109